Dryland Horticulture

Principles and Practices

THE AUTHORS

Dr Rajaneesh Singh is presently working as Professor and Head, Department of Horticulture, Tilak Dhari Post Graduate College, Jaunpur (U.P.). He did his Graduation in Agriculture from Udai Pratap Autonomous College, Varanasi, Post Graduation in Horticulture from G.B. Pant University of Agriculture and Technology, Pantnagar and Ph.D. in Horticulture from BHU, Varanasi.

Dr. Singh has served Tilak Dhari Post Graduate College in various capacities like In-charge Garden Section, Major in NCC, Member of the Sports Committee, Academic Counselor of IGNOU study center and Former Proctor of the college. Dr. Singh has 22 years teaching experience of horticulture. Dr. Singh decorated best teacher award in 2019 by Hon'ble Water Resource Minister Uttar Pradesh Government and Yashashvi Samman by Royal Association for Science Led Socio-Cultural Advancement in 2022. Dr. Singh has more than 89 publications to his credit (research papers 29, proceeding papers 8, conference/ seminar papers 16, books 7, book chapters 11 and popular/semi-scientific articles 18). He had been chairman of the organizing committee of One National Workshop. Dr. Singh qualified ICAR NET in Horticulture and Vegetable Science and received Senior Research Fellowship in ICFRE during Ph.D. programme. He is fellows of Indian Society of Vegetable Science and International Society for Noni Science. He has supervised 12 M.Sc. (Ag.) and 5 Ph.D. students as Chairman of Advisory Committee.

Dr Bijendra Kumar Singh is presently working as Assistant Professor, Department of Fruit Science, College of Horticulture, Banda University of Agriculture and Technology, Banda (U.P.). He did his Graduation in Agriculture from VBSPU, Jaunpur, Post Graduation in Horticulture from BBAU, Lucknow and Ph.D. in Horticulture from BHU, Varanasi. He has specialization in Fruit Science.

Dr. Singh started his career as an Assistant Professor, Horticulture at Tilak Dhari Post Graduate College, Jaunpur in 2017. Dr. Singh has 168 publications (research papers 36, proceeding papers 5, conference/seminar papers 55, books 7, book chapters 25 and popular/semi-scientific articles 40). He had been convener of the organizing committee of One National Workshop. Dr. Singh qualified ICAR NET in Fruit Science and received University Grant Commission (UGC) fellowship during his Ph.D. He is a life member of renowned societies and journals in India like Asian PGPR Society of Sustainable Agriculture, International College of Nutrition's, Society for Advancement of Research on Pomegranate, Trends in Biotechnology and Biological Sciences, Advances in Life Sciences and Trends in Biosciences, International Society for Noni Science, Indian Society of Horticulture Science and Progressive Horticulture. He has supervised 8 M.Sc. (Ag.) and 6 Ph.D. students as Chairman and Member of Advisory Committee.

Dryland Horticulture

Principles and Practices

As Per ICAR'S 5th Deans' Committee Syllabus

Dr Rajaneesh Singh

(Professor and Head)
Department of Horticulture
Tilak Dhari Post Graduate College, Jaunpur

Dr Bijendra Kumar Singh

(Assistant Professor)
Department of Fruit Science
College of Horticulture
Banda University of Agriculture and Technology, Banda

2023

Daya Publishing House®

A Division of

Astral International Pvt. Ltd.

New Delhi – 110 002

ISBN 9789359191157

Published by : **Daya Publishing House®**
A Division of
Astral International Pvt. Ltd.
– ISO 9001:2015 Certified Company –
4736/23, Ansari Road, Darya Ganj
New Delhi-110 002
Ph. 011-43549197, 23278134
E-mail: info@astralint.com
Website: www.astralint.com

World Noni Research Foundation

64, Third Cross Street, Second Main Road, Gandhi Nagar, Adyar, Chennai – 600 020
Telefax: 044-2442 3601 Email: mail@worldnoni.org Website: www.worldnoni.org

Dr. Kirti Singh
FNASc, FNAAS, FNABS

CHAIRPERSON

Foreword

India has made tremendous progress in the science of Horticulture. Presently Horticultural crops, covering nearly 7 per cent of the grass cropped area and contribute about 19 per cent of the gross value of India's agricultural output. Horticultural crops cover a wide range of flora comprising fruits, vegetables, ornamental crops, plantation crops, root and tuber crops, medicinal and aromatic plants, spices and condiments, if integrated in a land use system properly, will not only improve biological productivity, social and nutritional security, better export potential, high remuneration but also promote diversification of agriculture.

Till today major attention was given to food grains and resource. Now rich situation further expansion of area under horticultural crops on good land is least possible. The alternative only seems to be to utilize the complex diverse risk prone ecological situation like arid and semi-arid region. Where there is immense potential for production of quality fruits, vegetables, flowers and seed spice. There is also good scope for production of Horticultural crops. Therefore there is a need to intensify the research and development on under exploited crops of hot and cold arid ecosystem like Capparis, Salvadora, Cordia, Aegle, Feronia, Jamun, Karonda, Mulberry, Lasoda, Tamarind, *etc.* literature on this subject is limited. It is necessary to bring such publication.

The book entitled **Dryland Horticulture: Principles and Practices** shall bring out importance and scope of Dryland Horticulture, limitation and constraints of Dryland Horticulture and technology to overcome these problems through appropriate choice of these species, management skill, propagation techniques and water management. The book offer tremendous scope. Practices of twenty one fruits species based on Indian experiences which will be of significance use for students, teachers and those who are involve in research and development of Dryland Horticulture. Authors deserve all appreciations for their hard work and bringing out this useful document. I feel that they should be suitably awarded.

Kirti Singh

Former Vice-Chancellor

Preface

Traditionally fruit cultivation in India has been associated with fertile land orchard with assured irrigation. Since good irrigated land have already been used more or less completely. We have to think of large area of marginal lands in arid or semi arid regions of the country with little or no supplementary irrigation. The arid horticultural crops are also well prone to grow organically. There is also good scope for production of horticultural crops. In view of WTO regime of export requirement the peculiar dry climate and warm blend in hot arid region offers opportunity for producing quality product of high health standard.

The book will proved boon for student contesting Indian Council of Agricultural Research- JRF, SRF, ARS and NET in the field of horticulture. I am highly thankful to management of the college and principal for providing me always moral support in writing of this book. We have great pleasure in acknowledging moral support, encouragement valuable suggestion and connective criticism received from Dr. Hari Baksh, Mr. Raj Pandey, Mr. Aman Srivastava, Mr. Prince Mishra and all agriculture faculty members, Tilak Dhari P.G. College, Jaunpur.

I am highly grateful to Prof. Anil K. Singh (BHU), Prof. Ranveer Singh, Prof. J.P. Tiwari and Prof. Hari Har Ram who always encouraging me for preparing this manuscript. We appreciate the efforts of M/s. Astral Internaitonal Pvt Ltd, New Delhi for printing the book in the nice form.

Dr Rajaneesh Singh ***Dr Bijendra Kumar Singh***

Contents

Section-II
Cultivation Practices of Arid Fruit Crops

Section-I

Fundamentals of Dryland Horticulture

Chapter 1

Introduction

Arid climates/dry climates, are climates in which a deficiency of moisture restricts, but not necessarily inhibit, plant growth. Some would consider the term synonymous with desert climates, but others would claim the letter to be a special case involving high average monthly or annual temperatures. Attempt to define arid climates objectively have involved assessment of the precipitation effectiveness and evaporation is knowledge to be an important factor. Perennial fruit crops grown in all season or fruit crop which grows and yields without wasting any available rain water, nutrient from soil is selected and cultivated

The word 'Arid' originates from the Greek word 'Arere' meaning' to be dry'. The term "arid" normally means a region of the earth's surface where rainfall is nil or inadequate, with the result the vegetation is non-existent or sparse, agriculture difficult or impossible and human living conditions precarious.

It is a climate in which a deficiency of moisture restricts, but not necessarily inhibits plant growth. Water is always a limiting factor in this climate. Water deficiency is taken as the sum of the monthly differences between precipitation and potential evapotranspiration in the months receiving lower rainfall than the normal. On the basis of rainfall, it has been defined to be a scanty rainfall area with 0-250 mm rainfall/annum.

In arid zone/dry climate rainfall is very low and is continued to 2.0- 4.5 months and remaining dry months. The rains are also erratic and often come in a few big storms of short durations which result in great sum off losses instead of charging the soil profile resulting into soil as well as atmospheric water stress during major part of the year. Soils in dryland or arid areas are coarse and structure less with very low water holding capacity. Thus, dry regions have peculiar eco-climatological features and they can exist in tropics, subtropics as well as temperate zones of the world.

A humidity province within which value of precipitation effectiveness index which is equal to 10 times the sum of the monthly precipitation-evaporation ratio, are less than 16. On the moisture index, $\text{Moisture index} = 100 \times \frac{(\text{Surplus} - \text{Deficit})}{\text{PE}}$ scale, values of -60 to -40 to define an arid climate.

Suffice to say that water is limiting factor in arid climate. Aridity index is an indicator of water deficiency with respect to water need at any point. Water deficiency (d) is taken as the sum of the monthly differences between precipitation and potential evapo-transipiration in the below normal months. For the water deficient month, the sum of the monthly value of potential evapo-transpiration is given by n. then aridity index is = 100 d/n. on the basis of rainfall it has been defined to be a scanty rainfall area with 0 to 10 inches (0 to 250 mm) rainfall per annum. In fact in arid zone the rainfall is very low and is confined to 2 to 4.5 wet months and the remaining dry months. The rain are also erratic and often come in a few big storms of short durations which results in great runoff instead of charging the soil moisture profile resulting into soil as well as atmospheric water stress during major part of the year. The vapour pressure deficit range between 10 to 30 mb and sometimes exceed this limit during hot summer months. Soils of such area are coarse and structure less with very low water holding capacity. Thus arid region have peculiar eco-climatologically feature and they can exist in tropics, subtropics as well as temperate zone of the world.

Drought is a natural hazard, known primarily in terms of scarcity or supply deficit of water due to failure of rainfall, deficiency of reservoirs and water resources, resultant failure of crops, and consequently a level of serious socio-economic distress. Cultivation of crops in areas receiving rainfall below 750 mm is known as dryland horticulture. Dryland horticulture areas are grouped under semiarid regions. Therefore dryland horticulture is taken up in small land holdings and in water scarcity lands.

A fruit is a matured ovary with associated parts. In some fruits it is simple ovary whereas in other some parts are also involved like sepals, petals, stamens, receptacles and peduncles and develop at the same time as the ovary and form part of the fruit. Therefore from physiological point of view a fruit may be defined as "a structural entity that results from the development of the tissues that support the ovules of plant". A fruit in horticultural sense is something which is eaten fresh and out of hand or processed.

Fruit growing has been defined as the cultivation of edible fruits on woody plants. This definition excludes strawberry, gooseberry, *etc.* and would be even less appropriate for pineapple and banana which are not woody. Therefore, arid fruit culture or growing would be the cultivation of edible fruits in arid region that are consumed either fresh or processed.

Perennial fruit crops grown in all season or fruit crop which grows and yields without wasting any available rain water, nutrient from soil is selected and cultivated. Apart from this, it provides employment to farmer, women and youth. Fruit harvesting, packaging, grafting *etc.* tasks are carried out by village youths by developing as empowerment and obtain more profit. In our country, the fertile and irrigated lands have been utilized for production of food grain and commercial crops.

Important Arid Fruit Crops for different Arid Fruit Species of India

Sl. No.	Fruits	Botanical Name	Sl. No.	Fruits	Botanical Name
1	Ber	*Zyzyphus mauritiana*	13	Fig	*Ficus carica*
2	Pomegranate	*Punica granatum*	14	Phalsa	*Grewia subinaequalis*
3	Aonla	*Emblica officinalis*	15	Tamarind	*Tamarindus indica*
4	Custard Apple	*Annona squamosa*	16	Losoda	*Cordia myxa*
5	Guava	*Psidium guajava*	17	Jamun	*Syzygium cumnii*
6	Date Palm	*Phoenix dactylifera*	18	Karonda	*Carissa carandas*
7	Bael	*Aegle marmelos*	19	Rose Apple	*Eugenia jambos*
8	Wood Apple	*Feronia limonia*	20	Lakooch	*Artocarpus lakoocha*
9	Mulberry	*Morus spp.*	21	Carambola	*Averrhoa carambola*
10	Kher/Kair	*Capparis apphylla/ Capparis decidua*	22	Barbados Cherry/West Indian Cherry	*Malpighia glabra*
11	Citrus	*Citrus spp.*	23	Mangosteen	*Garcinia mangostana*
12	Karonda	*Carissa carandas*	24	Hingot	*Balanites aegyptiaca*
13	Gugal	*Commifera wightii*	25	Pilu	*Salvadora oleoides*

Scope of Arid Fruit Culture

About $1/_3$rd land surface of the world is in the grip of arid zone and hot deserts. The man made desert alone covers 910 m ha area. The scope of the growth of dryland fruit industry shall be determined by:

- **a)** Incentives for farmers
- **b)** Necessity
- **c)** Adaptability of the crops
- **d)** Future scope for expansion depending on the availability of inputs, infrastructure for distribution.
- **e)** Marketing system and industrial support.

Scope of growth of arid fruit industry shall be determined by incentive for farmers, necessity, adaptability of the crops and future scope for expansion depending on the availability of inputs, infrastructure for distribution, marketing system and industrial support.

1. The importance of fruits in providing valuable nutrients, particularly vitamins and minerals in the human diet is well known and according to dietary standard an individual must consume 120 g fruits/capita/day. According to present 102 MT scenario of total fruit production in India and out of that 10 to 12 per cent as post-harvest losses, 7 per cent as consume in processing purpose and 5-8 per cent as consume in export of the total fruits production. The present availability of fruits in India is 203 g/capita/day. To bridge this gap of demand and supply the production will have to be raised through increasing orchard efficiency and by bringing more area under fruit crops.

   The average yield of fruits in India is around 14 tonnes/ha. Yields in most parts of India are well below the proven potentials, results of research prove that fruit yields can be raised 2 to 6 times than the national average by planting the best available varieties and using improved orchard management techniques. Traditionally fruit culture has been associated with fertile garden lands with assured irrigation. Since good irrigated lands have already been used, more or less completely, for raising high yielding cereals, commercial crops and horticultural crops and even for ever increasing industries and housing, we have to think of the large areas of marginal lands in arid and semiarid regions of the country with little or no supplementary irrigation. Therefore, fruit culture in arid region is the only hope of the future for feeding the fast increasing population with protective food like fruits.

2. At present most of the arid zones of India has animal husbandry based economy which is not sufficient to provide the required stability in the income of people and therefore, people at large in this region are nomadic in habits. Provision of sustained and stable income to these people shall be sufficient incentive for them to grow fruits. We are now aware that there are a number of fruit crops which have great capacity to stand drought and salinity and still provide good income to growers. For example, the ber which was considered to be a wild fruit a decade back has gained commercial importance in the states of Rajasthan, Gujarat, Haryana and Punjab.

3. These regions are characterized by conditions of high aridity. The atmospheric humidity is very low. The rainfall is not only low but erratic and confined to a period from July to September. Summer months are dusty and hot which would not permit proper fruit development. Therefore, the strategy of fruit growing should be such that fruit species selected should complete their reproductive phase well before such stress period and trees should have xerophytic features to overcome drought problem. Plants with such adaptabilities are known and available like ber, lasoda, kair, pilu and they have great food value as well as industrial uses if only properly exploited.

5. **Farm ponds:** Construct the farm pond near valley area. Initial months, the collected water are used for irrigating. Construct the rim of bund with more height than land surface.
6. **Avoid wind blow:** Grow tall trees in North-South direction to avoid wind blow damage. The tree species also helps in retaining moisture in the field (avoid evaporation). Best results are observed when the trees are planted in triangular shape.
7. **Coverage:** Dry grass, dried leaves are covered on soil to avoid water evaporation, weeds growth and soil erosion. Now a day black polythene sheets are used for coverage.
8. **Use of non evapourating materials:** Some chemical materials sprayed on leaves mainly slow down in evaporation process by formation thin sheet layer. Various materials are in use presently, *e.g.*, Acropol is used in grapes, Polycarp is used in banana crops *etc.*
9. **Drip irrigation:** In this irrigation system about 30-50 per cent of water is conserved. The well developed drip irrigation system, the equal amount of water of 5-10 liters is supplied to each tree crop at same interval of time from small drips closed with cotton.

Present Status of Arid Fruit Culture in India

The Indian hot arid zone covers an area of 31.7 m ha. (12 per cent of country's total geographical area) in states *viz.*, Rajasthan (61 per cent), Gujarat (20 per cent), Andhra Pradesh and Karnataka (10 per cent), and Punjab and Haryana (9 per cent). The remaining 7 million ha is a cold arid zone situated in the Ladhakh area of Jammu and Kashmir and the Lahaul Spiti area of Himachal Pradesh. The hot arid zone lying west of Rajasthan occupies the largest area (20 million ha) of the arid ecosystem of the country, and is popularly known as Thar Desert.

It is amply clear from the table that major arid zone lies in north-west region of the country. These areas are inhabited by about 20 million people and over 23 million livestock population being the most thickly populated arid zone of the world where severe biotic exploitation is occurring in the face of a very limited resource base.

Economy of this region is mainly based on animal husbandry not sufficient to provide required stability and therefore major part of population leads a nomadic life. Crop productivity is erratic and beset with uncertainty and high risk factors. The common annual crops are pearl millet (bajra), mung bean, moth bean, cluster bean and cow pea. The perennial pasture grasses are *Lasiurus sindicus* (Sevan), *Cenchrus ciliaris* (Dhamen), *C. setigerus, Dicanthium annulatum, Chrysopogon fulvus, Panicum antidotale, Prosopis cineraria* (Khejri), *Prosopis juliflora* (Vilayati babool), *Acacia senegal* (Kumat), *Salvadora oleoides* (Pilu), *Tamrix articulata* (Farash) and *Tecomella undulata* (Rohida) among tree species, *Ziziphus nummul*aria (Jharber), *Calotropis procera* (ak), *Capparis decidua* (Kair), *Calligonum polygonoides* (Phog) and *Acacia jacquemontii* among shrubs provide vegetation cover in these regions. Pasture legumes like *Clitoria ternatea, Dolichos lablab* var lignosus, *Stylosanthes hamata, S. scabra* and Siratro, tree species like *Acacia tortilis, Albizia lebbak, Azadirachta indica, Cassia siamea, Holoptelia intigrifolia, Dichrostachys glomerata, Colophospermum mopane, Leucaena leucocephala, Sesbania aegyptiaca, Dalbergia sissoo* and shrubs like *Euphorbia antisyphilitia, Saccharum bengalense, Simmondsia chinensis* (Jojoba).

In these areas wherever assured irrigation is available through canals, wells, ponds *etc.*, fruits like pomegranate, guava, lime, sweet oranges, 'Kinnow' mandarins, grapes and papaya are grown. However, under rainfed conditions very little fruits are grown inspite of tremendous potential for tree crops like ber, Lehsua, Aonla, Mulberry, Karonda, Hingot, Gugal, jojoba, *etc.*

The available statistics reveal that area under fruit crops is dismally low resulting in inadequate supply of this commodity leading to malnutrition inspite of tremendous potentialities. It has been probably due to ignorance as fruit culture traditionally has been associated with fertile garden lands with assured irrigation and therefore, scientists and farmers have paid no attention to this rich resource. There is a great need to increase local production of fruits in these dry areas to improve the economic conditions of the people, their nutrition and health standards and to provide stability in life.

Planting Techniques in Arid/Dryland Orcharding

a) **Selection of crops:** Selection of suitable horticultural crops depends upon different agro-climatic conditions in Karnataka 3 categories of areas are prevailing.
   - **i.** Maidan- The crops are jack fruit, tamarind, annona, ber, jamun, wood apple amla, fig, phalsa, *etc.*,
   - **ii.** Malnad– Early varieties of ramphal, sitaphal, jamun, jack fruit *etc.*
   - **iii.** Coastal– Early varieties of karonda, jamun, jack fruit, bread fruit, butter fruit, cashew, apple *etc.*

b) **Wind breaks:** like casuarinas, silver oak help in demarcating the boundaries and also provides timber.

c) **System of planting:** Depending upon the level and slope of the land system of planting square, triangular, rectangular or contour systems of planting could be adopted.

d) **Planting of rootstocks and in-site grafting:** In rainfed areas it is always desirable to plant seedlings and to take up in situation grafting in subsequent season, helps in better establishment.

The unexploited fruit plants for watershed areas are tamarind, ber, karonda, pomegranate, jamun, jack fruit, aonla, fig, phalsa, wood apple, custard apple *etc.*

Distribution of Arid and Semiarid/Dryland Regions in India

The climatological data of the arid and semi arid regions of India. The entire dry farming areas have been divided into four zones, Zone –D, E, F and G.

- ☆ **Zone-D:** This zone has the lowest crop potential. It occurs at three places. The first starts from the Jamnagar district of Gujarat to Ferozpur in Punjab. The second includes part of Ahmadnagar, Pune, Satar, Solapur and Sangli districts of Maharashtra. The third includes portions of Bijapur, Raichur and Bellary districts of Karnataka and Karnool and Anantpur districts of Andhra Pradesh.
- ☆ **Zone-E:** This area extends from Rajkot in Gujarat along with east of Zone-D up to Punjab through Rajasthan and Haryana. This area also extends from Ahamadnagar (Maharashtra) to coastal areas of Cadappah (Andhra Pradesh) through Stara, Pune, Solapur and Sangli districts of Maharastra and Bijapur, Bellary and Tumkur districts of Karnataka.
- ☆ **Zone-F:** This zone is confied to two localities in the country. In the north, it comprises vast areas of Gujarat, Rajasthan, Uttar Pradesh and a portion of Haryana and Punjab. In the South, this zone stretches from the Nasik district of Maharashtra to Kanyakumari of about 1/3rd land surface of world is in the grip of arid zone and hot desserts. The man made desert alone covers 910m ha area Tamil Nadu. A large portion of Karnataka and Andhra Pradesh also fall within this zone.
- ☆ **Zone G:** This zone has the highest crop potential. This area consists of small portion of Uttar Pradesh, Madhya Pradesh, Gujarat and Tamil Nadu and considerable parts of Maharashtra and Andhra Pradesh. A part of Tamil Nadu receives the North east monsoon and the growing season, therefore differs significantly from the rest of localities.

Major Problems and Constraints in Fruit Productivity in Arid Region

Interest in fruit culture in the arid regions remained sporadic and as a hobby just for the satisfaction of the aesthetic urge of the royal families and the cities who had all resources at their disposal, It could never become a profession. However, wherever the fruit plantations occur, they are often raised with ingenious techniques and concerted efforts of a few. Of course this lack of interest has some major reasons and constraints. The major ones are:

1. Rainfall in hot arid zone is very low and is confined to the period from July to September, with 19 to 21 rainy days out of 12 to 31 days in the whole year, resulting both in soil moisture stress and atmosphere water stress to the plants after the rainy season. After April, the vapour pressure deficit is more than 24 mb and exceeds 30 mb during May and June. To be precise, water is a limiting factor. The wind velocity is generally very high aggravating the problem.

2. High solar radiation incidence (450 to 500 cal./cm^2/day)And high wind velocity (20 km/hr) result in a high potential evapo-transpiration (6 mm/ day) and high mean aridity index (74 to 78 per cent).
3. The soils are of poor quality with low fertility level and poor water holding capacity (25 to 28 per cent) and with high infiltration rate (9 cm/hr). The soils are generally shallow with sub soil under lain with carbonaceous concretions which has variable thickness. In fact, there are hardly Class I and Class II lands. Class VI. Class Ill and Class IV lands, in that order, occupy the major areas.
4. The ground water resources are scarce and at great depths (40 to 50 m) and mostly saline in nature due to poor drainage and recharge.
5. Lack of interest in exploitation of plants, amenable to arid conditions like ber, Salvadora *etc.*
6. Since the arid regions have typical ecological features, special technology is required for success in fruit production. This technology with respect to suitable plant types and cultural practices including cropping systems has not been properly developed to win the confidence of farmers.

Present Research Set-up on Arid Fruit Culture

Fruits have always been considered as luxury in our country and therefore, skipped the attentions of planners and more so In case of arid fruits which remained neglected even by researchers who gave priority to high value fruits like mango, banana, Grapes, citrus and apple. A beginning was probably made by Central Arid Zone Research Institute, Jodhpur (CAZRI) by Working on water harvesting and moisture conservation, fruit fly Resistance and rootstock studies in ber, improvement in pomegranate for quality attributes (aril colour, sweetness and seed content); evaluation of germplasm, control of Graphiola disease and processing techniques in datepalm. Some work on nutritional studies and irrigation scheduling in datepalm, propagation in *Capparis decidua*, germplasm collection in Annona has Now been initiated. Institute has its sub-stations at Pali and Bikaner (Rajasthan). The Indian Institute of Horticultural Research, Bangalore has done some work on chemotaxonomy of Annonaceous fruits and pomegranate evaluation. Its experiment station at Godhra has started work on horti-pastoral system and Silvi-horti-Pastoral system for arid regions. The National Bureau of Plant Genetic Resources, Regional Station, Jodhpur conducts some work on germplasm survey and Conservation for arid zone fruit crops. Punjab Agricultural University, Ludhiana; Haryana Agri-Cultural University, Hissar; Rajasthan Agricultural University, Bikaner and Gujarat Agricultural University, S.K. Nagar also Work on fruits like ber, datepalm, pomegranate *etc.* Real impetus on research on arid zone fruits was started Through Agriculture Production Cess Fund Scheme entitled,

Research on some selected fruits in the semi-arid areas in India at 10 centres during the period (1.7.76 to 30.7.81) spread over the states of Punjab, Haryana,

Himachal Pradesh, Rajasthan, Gujarat, Maharashtra, Karnataka and Tamil Nadu on fruits like datepalm, ber, pomegranate, fig, custard apple, walnut and apricot. This scheme was merged to create Cell III of the All India Coordinated Fruit Improvement Project (AICFIP). During the Seventh Five Year Plan, the Cell III of the AICFIP was restructured and strengthened to form an independent All India Coordinated Research Project for Arid Zone Fruits (AICRPAZF) with 13 centres. Under this cell coordinated with head quarters at H.A.U., Hisar, Coordinated Research for testing technologies with respect to her, pomegranate, datepalm, fig, custard apple, aonla, and bael is in progress. Finally a National Research Centre for Arid Fruits is in making. Progress of work on these crops is being reported in the form of proceedings of workshop every year and the scientists also bring out publications through journals of different horticulture societies like Haryana, Rajasthan, south Indian horticulture society *etc.*

To stabilize production and win the confidence of orchardists in arid-region, work on following aspects is a priority.

1. Survey, collection and evaluation of large number of Indigenous plant types of economic significance wild in arid Regions of the country with special emphasis on drought and Salinity resistance as well as quality attributes and conservation Thereof.
2. Improvement work on ber, pomegranate, wood apple, bael, aonla, *capparis decidua*, Salvadora *etc.*, which could not be given the required attention.
3. Maintenance of a directory of lands and locations under marginal situations such as drylands, semi-deserts, scrubs, bald hills, hilly terrains *etc.*, with full information about their agro-ecological features. Then evolution of drought resistant crops which perform well under moisture stress conditions and car-marking them for various situations in arid region.
4. Generation of information on soil and water status for optimizing water utilization in arid zone for fruit crops.
5. Development of efficient water use technology for increased production under both rained and irrigated conditions in the arid regions.
6. Generation of knowledge about the understanding of physiological adaptations for salt resistance in fruit crops and identification of salt tolerant genotypes as salinity and sodicity of soil and water are typical features of arid region.
7. Development of fruit tree based cropping system to provide stability in income to people in arid regions.
8. Development of effective plant protection measures against powdery mildew and Isariopsis black spot diseases of ber, rust in aonla, Graphiola leaf spot in datepalm and leaf and Fruit spot in pomegranate and insect pests like ber fruit fly, and Pomegranate butterfly which have become constraints in fruit crop production in this area.

9. Standardization of techniques for faster multiplication of planting material and establishment of nurseries to raise and Supply drought tolerant fruit crop plants to prospective growers.
10. Developing suitable technologies for reducing post harvest losses and supporting cottage industries based on arid Fruits.
11. Strengthening of extension activities to educate farmers on fruit culture in arid region and financial assistance on easy term for fruit culture.

Chapter 2

Agro-Climatic Zone

Arid zones are characterized by low and high variable precipitation compare to high evaporative demand. Therefore moisture stress is the characteristics feature of the arid zone. Arid region in India occupy nearly 12 per cent of the total area sustaining a human population of about 20 million and livestock 23 million. Being one of the most thickly population among the desert area of the world. Recurrent drought leads to disappear to famine and helplessness for the inheritance of the region and off course great nation liability. Basically arid zones are classified into two subzones.

1. Hot arid zone
2. Cold arid zone

1. Hot Arid Zone

Hot arid zone spread in 0.32 million km. In Rajasthan 62 per cent, Gujarat 20 per cent, Andhra Pradesh 7 per cent, Punjab 5 per cent, Haryana 4 per cent, Karnataka 3 per cent and Maharashtra 4 per cent are covered by hot arid zone. This is almost 90 per cent of in north western part of the country.

The precipitation of scanty and erratic (30-500 mm), solar radiation is very high (450- 500 calories/cm^2 per day) and wind velocity is also high 20 km per hour. The rain is confined from July-September with 09-21 rain spell out of 12-30 in whole year after April. The vapour pressure is deficit is more than 24 mb and it is 30 mb during may June. The condition course intense Evapo-transpiration 6 mm/day increase means aridity index 74-78 per cent.

Soil being wind blonde 85 per cent sand with low moisture (0.1-0.45 per cent) poor water holding capacity (25-28 per cent) and high infiltration rate 9 cm/hour. The ground water resources are very less at a great depth (40-50 per cent) having poor surface and subsurface drainage. The ground water a large part of the arid area. Ex. 60 per cent of the arid Rajasthan arid highly saline. The soil are generally swallow bellow the soil depth of 90-120 cm. they exist the prone zones having carboneous concretion which have variable thickness.

2. Cold Arid Zone

The cold arid zone occupied 0.107 million square km. west of Himalayas mainly Laddakh (Leh) and Karagil in Kashmir, Lahaul spiti in Himachal Pradesh and Gangotri and Yamunotri in Uttrakhand. The cold regions are the hilly tracts (300-5000 m) above mean sea level. The border area of Tibbat and Nepal the all problems of high altitude prevalence scanty rain (100-200 mm) with heavy snowfall, high wind velocity, poor soil fertility and low temperature (-9 to -14 °C) particularly in winter (December-February).

General Climatic Features of the Arid Zone

1. Winter Season (December to February)

During winter the sky remains clear in north western arid region. Low relative humidity large diurnal (Day-night) variation of temperature and light north easterly winds are the feature of the weather. These condition changes only when western disturbances pass over the North West India resulting into cloudy weather and rain. The winter rainfall is 7-12 per cent in arid zone of Punjab and Haryana 3-6 per cent in Rajasthan and negligible in other areas. Occasionally frost in North West arid region due to cold wave on the rear side. The frequency of frost in such regions is once in three year. The normal minimum temperatures during the coldest month in different regions are being presented in table below.

Normal Minimum Temperature during the Coldest Months in different Arid Regions

Region (State)	*Normal minimum Temperature with range*	*Coldest Month*
Ladakh	-9 to -14 °C	January
Punjab and Haryana	3 to 5 °C	January
Rajasthan	3 to 10 °C	January
Gujarat	11 to 15 °C	January
Maharastra	11 to 15 °C	December
Karnataka	11 to 15 °C	December
Andhra Pradesh	16 to 18 °C	December

The winter season is very dry excepting in coastal areas the mean vapour pressure deficit value are less than 10 mb. The relative humidity is generally high in morning due to very low air temperature.

2. Hot Season (March to May)

From March onward temperature starts increasing in faster rate. May and June are the hottest month in North west India. During these month dry and hot dust raising wind occur very frequently which occur more frequently during the year preceded by very low rainfall and cause enormous wind erosion. The mean maximum temperature in different arid region during month are as fallows.

Normal Maximum Temperature during the Hottest Months in different Arid Regions

Arid Region (State)	*Range of Normal Maximum Temperature*	*Hottest Month*
Ladakh	25 to 31 °C	July
Punjab and Haryana	41 to 42 °C	May
Rajasthan	41 to 42 °C	May
Gujarat	32 to 41 °C	May
Maharastra	40 to 41 °C	May
Karnataka	39 to 41 °C	May
Andhra Pradesh	40 to 41 °C	May

Monsoon Season (June to September)

The diurnal variations in the temperature are minimum during monsoon season. The daily maximum temperatures in North eastern area are in the order of 35-38 °C. Humidity is higher but high strong winds are common resulting into high evaporative demand for water. However the period is very congenial for most plant growth.

Post-monsoon Season (October to November)

After the withdrawal of south west monsoon dry whether prevails till the end of October in North West India. However in southern arid zone rainfall occurs due to north east monsoon. Now concede the significance of each phenomenon like rainfall, light, heat, water (Moisture and relative humidity) and air.

Distribution of Rainfall

On the basis of annual precipitation the regions can be grouped as follows:

1. **Arid region:** In arid region where rainfall are not more than 250 mm (10 inch).
2. **Semi arid region:** Rainfall is 250-625 mm (10-25 inch). The mean annual rainfall varies from 100 mm in North western sector of Jaisalmer District to 450 mm in the Eastern boundary of arid zone of Rajasthan. It varies from less than 30 mm to 500 mm in arid zone of Gujarat. And from 200-450 mm in Haryana and Punjab region.

The seasonal distribution of rainfall over the country is that a major portion of annual rainfall is confined to south west monsoon season from June to September in the arid tract of south India. The south west monsoon rain commence by first week of June while North Indian arid zone by the beginning of July. Whole of Gujarat and South and central Rajasthan receive the 90-95 per cent rainfall from south west monsoon. The contribution of winter rainfall range from 9-12 per cent of the annual value in addition to 78 per cent contribution from south west monsoon season in Punjab.

In Karnataka and Andhra Pradesh the contribution of south west monsoon in the order of 48 to 67 per cent and north eastern monsoon contribute 21-32 per cent of the annual rainfall. It means the major rainfall is concentrate only to three month (July to September) and rest time is dry period.

Light: Light is available in plenty in arid region of India. Therefore depending on the latitudinal situation draught tolerant crop can be selected. Day length exerts considerable effect of photosensitive plants. However fruit plants in general show only slight reaction are none at all to the photoperiod.

Heat: Growth rate of the plant depend primarily on temperature depending the location. The range varies from Ladakh where normal minimum temperature in winter range from (-9 to -14 °C) in January in other places hardy fruit crops would have no problem of growth and development. The growth rate of the plant depends primarily on temperature. The heat index is a term which indicates the sum of average daily temperature during the growth cycle of the crop. For perennial crops however average temperature a not a good criteria as the plant stop growing when it is too cold or too hot. It is therefore bitter to use only effective temperature those between two extremes for the growth. Extremely high temperature cause wilting, sunscald, narcotic spot and even the death of the plant. The date palm is resistance up to 50 °C and so is the case with Ber, Kair and Salvadora. The minimum and maximum temperature is support plant growth lies between 4.5-36 °C.

Water: Atmospheric humidity and soil moisture together constitute another factor of climate having profound influence of plant growth. The soil moisture is dependent not only on rain distribution but also on rain intensity. That is precipitation per unit time. It is not the total rainfall but distribution is more important. The number of dry month in which evapo-transpiration exceed precipitation is of great significance because it is on this criteria that creation of irrigation facility and choice of crop would depend

Water is also present in atmosphere as vapour. Atmospheric humidity influence growth and development of the plant. Low humidity has a drying effect while high humidity create favorable condition for fungi for causing diseases. This will influence planning on plant protection and atmospheric humidity may influence setting also. High humidity is good for setting in custard apple and Ber while low humidity in Date palm. Generally 50 per cent relative humidity is optimum in atmosphere for good plant health and growth.

Air: Air is usually not a limiting factor for plant growth. However under high intensity of life carbon dioxide may become may become a limiting factor. Wind is an important consideration at it can cause great damage to the crop. Storm 50 km/hour, hurricanes 100 km/hour or above. May up-root tree caused fruit drop. Adversely affect the fruit set and increase evapo-transpirational loss. Air pollution is dangerous to the plant as to the human being. Therefore the quality of air becoming important in plant growth. The important pollutants are Ozone (O_3) per oxy acetyl nitrate (PAN), Nitrous nitrogen oxide, sulfur dioxide, fluoride, ethylene and soil particles. Sedimentation is typical to arid condition which may limit photo syntheses to plant.

Chapter 3

Arid Zone Soils of India

Soil is a reservoir of nutrients and moisture and also provides anchorage to plants and therefore, intimately associated to plant and its development. Since the plants live on this crust of earth, an understanding about the phenomenon is necessary to be wise on decisions. It is defined as a three phase System in which plants grow. All these three phases solid, liquid and gas are essential. Therefore, it is far more than an inorganic mass of debris-it is a biological system in the state of dynamic equilibrium. Solid part of the soil is rather a frame which provides space for other two phases liquid and gas. It consists of minerals, clay minerals, and organic matter which interact with each other, and, land properties to soil which can be grouped into two (i) physical properties and (ii) chemical properties.

Physical properties of the soil are the functions of level of organic matter, its texture and structure which are most significant aspects in crop production. Soil is a living system and among living beings earthworm, the bacteria, fungi and other microbes constitute the bulk of soil organism. They also contribute to organic matter besides the decomposed plant and animal material. Such materials are attacked by soil microbes and converted into simple materials; carbohydrates to carbon dioxide; proteins into amino acids and then to nitrates under the influence of nitrifying bacteria and the unbroken part is referred as humus which affects the soil structure and availability of nutrients. Depending on the level of soil organic matter the soils are again classed as mineral soils (organic matter less than 20 per cent) and organic soils (organic matter more than 20 per cent).

Texture refers to the size of soil particles. Mineral substances of soil are consisting of different size of particles: stone, gravel, coarse and fine sand, silt and clay. Proportions of the particles determine the texture of soil and accordingly there is textural classification of soil in which there are twelve classes Among these particles, clay is most important component on which cation exchange capacity of soil depends and proportion of different particles also determine water holding capacity of the soil.

Soil structure refers to the arrangement of soil particles into aggregates. A structure which assures sufficient quantity of water and air for root growth is considered to be good and therefore, crumb structure is preferred. Good structure also means stability. Water holding capacity of soil is dependent on its structure and therefore, important horticultural.

Chemical properties of soil determine the availability of nutrients to plants. These properties are cation exchange capacity (CEC), soil reaction *i.e.* pH, soil fertility, salinity and alkalinity.

Cation exchange is the most significant feature of soil colloidal particles like clay and humus; they have capacity for attracting cations, being negatively charged, and subsequent exchange of one ion from another a process referred as cation exchange. The process does not allow the loss of nutrients through leaching and makes them available to plant by exchange with H-ions on root surface. Cation exchange capacity of soil is variable which is expressed in milli-equivalents (meq) per 100 g and it is equivalent to the milligrams of H that will combine with 100 g of dry soil. This varies with soil type:

Humus	: 150 to 300 meq/100Dg
Montmorillonite	: 100 meq/100Dg
Kaolite	: 10 meq/100Dg

Soil reaction is referred to acidity and alkalinity of soil expressed as soil pH, which is the logarithm of reciprocal of the H^+ ion concentration and usually expressed in units from 0 to 14. Availability of nutrients and behavior of soil microbes are greatly influenced by soil pH. Proper pH for fruit tree growth is 6 to 7. Abnormally high, more than 9, and low pH less than 4, are toxic to plants.

Soil fertility refers to the level of nutrients in the soil, which is only indirectly related to the chemical composition of the primary inorganic minerals. The most important factor in soil fertility is the level of the forms of the nutrients available to the plants. These levels are under the influence of many factors including chemical and physical properties of soil.

Salinity refers to salt concentration in soil. A soil with Electrical Conductivity (EC) more than 4 mmhos/cm^2 pH less than 8.5 and Exchangeable Sodium Percentage (ESP) less than 15 per cent is considered to be saline and availability of nutrients is greatly influenced by it. It affects plants by raising the osmotic pressure of the medium surrounding the roots as well as by specific ion effect. Among various salts, chlorides are more toxic.

Alkalinity refers to the soil with Exchangeable Sodium Percentage (ESP) more than 15 per cent, conductivity less than 4mmhos/cm^2 and pH between 8.5 to 10. Soil with ESP less than 7.5 can be put for use in fruit cultivation, at 15 ESP it needs careful management and above this is unfit unless there are resistant crops available.

Soil moisture is the liquid phase of soil which is important to plants as constituent and carrier of nutrients. Maximum amount of moisture that is retained after the surface water is drained and after the water that passes out of the soil by gravity is removed is referred as Field Capacity (FC). The moisture content at which irreversible wilting occurs is known as Permanent Wilting Point (PWP) and the available moisture is the difference between the field capacity and permanent wilting point (FCPWP). Soils differ in their capacity to hold water depending on soil structure and texture. The medium textured soil which has more water available should be most suited for fruit crops.

In general it is believed that fruit crops should be grown on soils with good physical properties, in general loamy sand or sandy loam is preferable. Chemical fertility is desirable but not essential. In most cases the pH should be near 6.0. In fact, in order to serve as a medium of plant growth, the soil needs to have (a) sufficient depth to give a foot hold to the plant, (b) moisture storage-supply capacity, (c) a balanced supply of a variety of nutrient elements. In addition the soil should be well aerated and devoid of toxic concentration of salts. In comparison to annual field crops, the raising of fruit crops creates a several special situations are: (1) They need greater oil depth, (2) for equivalent green biomass the feeding zone is more confined than in case of annual crops, (3) the requirement of moisture and nutrients is prolonged and episodic. It would be desirable to interpret the arid region soils in respect of these points.

Brief Description of Major Arid Region Soils

In arid Rajasthan, dune soils are dominant in 30.6 per cent of the area and a subdominant. Associate in another 34 per Cent. These are fine sandy, light yellowish brown, structure less, Weekly to moderately calcareous (non-calcareous in the north- east). The light brown sandy soils occupy nearly 40 per cent of the area. These are loamy fine sandy in texture, structured but variable in calcareousness and in depth and degree of development of lime concretionary strata. Light textured soils with a hard pan at 40 to 60 cm soil depth constituted 6 per cent. The Grey Brown loams occupy another 13.6 per cent. These are mainly loam to clay loam with well developed blocky structure and with a strongly developed Weekly lime concretions strata.

In Punjab and Haryana, arid zone is made up of light to medium textured Indo-gangetic alluvial plain soils, varyingly covered with local and aeolian sands in the form of sheets and dunes. The soils are yellowish brown, non-calcareous to moderately calcareous. The soils of Ghaggar and Chautang river courses distinguish by their high percentage of silt and clay.

In Gujarat, the areas neighbouring Barmer and Jalore are dominated by sandy and locally saline soils. But further south in Kutch peninsula, the pediment slopes are covered by brown to dark brown clayey soils with a prismatic structure in the Sub-soil. The pediment slopes have a yellowish brown, medium textured, often highly calcareous soil. Coastal soils have at places a compact, sodic sub-soil.

In Andhra Pradesh and Karnataka, the area in Anantpur is dominated by recd sandy soil while in Kurnool, it is mixed and black soil. Similarly in Raichur district, the soil is from mixed black, deep black and red sandy type, while in Bellary it is predominantly red sandy type. Black soils in low rainfall areas have low organic matter, NP and available Fe and Zn, while red soils are light textured, shallow to medium in depth usually underlain with compact sub soils which allow free movement of water but impede root penetration. These soils are highly erodable and prone to crusting. Two major limitations for crop production on red soils are low fertility and moisture retention that most of the soils are not only poor in water holding capacity but also poor in fertility. The soils are coarse textured sandy and almost structure less with high infiltration of water. There are hardly any Class I and Class II lands. Class VI, Class III and Class IV lands, in that order, occupy the major area. They are often underlain with murram at varying depths.

Soil Characteristics in Relation to Fruit Crops

Fruit crops in general require greater soil depth. Their feeding zone is more confined compared to annual crops for equivalent biomass production and their requirement of moisture and nutrients is not only prolonged but also more episodic. Therefore, examined the arid soil characters in context with fruit growing as under:

Soil Depth

Most of the arid zone soils are underlain with a lime concretionary layer (Murram) at depths ranging from 60 to 150 cm 30 However, there is great variation in the frequency as degree of in duration. In extreme situations these are highly indurated and rock like forming perfect barrier through which moisture may find its way, though slowly but roots are inhibited. Sometime this strata serves as a moisture supplying zone during the extended period when the solum proper has lost all of its moisture.

Soil Moisture

Light textured soils dominate the soil cover in arid zone of India. These soils can retain only 50 to 120 mm of water in the solum which is very low. This low moisture retention capacity is because of the fact that pore space is primarily made up of non-capillary sizes which cannot hold water against gravity. Therefore, any water supplied excess of unfilled storage capacity goes as a deep percolation loss.

Another important consideration is that the movement of Water from a point of lower tension to the one at higher tension, say root surface, is very low due to lack of continuity in capillary space present. Thus these soils, whereas have, a rapid infiltration and high saturated state hydraulic conductivity, in unsaturated condition, which is normally found in field condition, is very low. Therefore, utilization of stored water solely depends on root system and the plants with greater net work of roots are at advantage. However, situation calls for light and frequent irrigation.

Soil Fertility

Most of the arid zone soils are well supplied with various nutrient elements. Only exception is nitrogen, iron, manganese and zinc. However, since the root zone of fruit crops is more confined, they show hunger signs. Localized applications of manures and fertilizers will require due attention from scientists.

At the end it can be epitomized that fruit crops require soil with good physical properties and, chemical fertility is desirable but not essential. However, in arid zone of India, soils are very poor not only in water holding capacity but are also poor in fertility. The soils are coarse textured sandy and almost structure less with high infiltration of water and often underlain with murram in varying depth. For fruit cultivation they will require good management, and call for developing technology for localized applications of nutrients and ameliorates. At the same time, appropriate selection of plant species will have to be made to suit to these situations

Chapter 4

Water Management

Water is the most important input for an assured fruit crop production programme. In the times of scarcity supplementary irrigation helps in overcoming production loses. However, context with arid region, this is scarcely available input and in case of its availability at any level, it requires to be judiciously used. In many arid and semi-arid areas underground water the only source of irrigation during non-rainy periods. From the practical point of view there are three major aspects of water management- (1) when to irrigate, (2) how much to irrigate and (3) how best to irrigate. They would demand careful considerations.

Under normal conditions, it is believed in most fruit crops, irrigation is best applied when the water tension in active root zone goes above 4 atmospheres or when 60 per cent of the available water in root zone is depleted. However, under arid conditions, where water is a limiting factor, we should identify moisture sensitive periods of crop and see that plants do not suffer during critical stages of growth like new flush, blooming, and fruit set, seed development and fruit growth and these should act as determinants for when to irrigate than only soil moisture tension values.

For determining, how much to irrigate, the principal is to apply sufficient quantity of water to bring the soil up to field capacity at a depth and area commensuration with bulk of the feeder zone. To achieve this objective, the method of irrigation becomes a point of significance *i.e.* how best to irrigate. Now when the water is finally flowing in the channels to the farm, irrigation may take place in two ways; (1) by gravity (surface irrigation) and (2) under pressure. There is another classification about the system of irrigation: Surface irrigation, sub-irrigation, sprinkler irrigation, and drip or trickle irrigation. Gain within each method there could be various methods to attribution like under surface irrigation, water could be distributed, by flooding, border method, basin flooding, furrow irrigation, ring irrigation *etc.* However, for arid conditions we have to make a choice for most efficient method and automatically it goes to drip irrigation, being most befitting technique, which shall be dealt in detail.

Drip irrigation is known by various names like trickle irrigation, or high frequency irrigation or daily flow irrigation. This is a method of watering plants at a rate equivalent to its consumptive use so that plants would not experience any stress through growing phase. In this the water is conveyed from Source (well, pond or tank) under low pressure to the root zone of the crop only. The twin objectives of this method of irrigation are: (1) provision of optimum quantity of water to the crop for optimum production and (2) saving the valuable water from wastage thereby increasing water use efficiency and the command area.

A typical drip irrigation system has the following components: (1) a water supply pump at the source of water, (2) Tilters, fertilizer mixing tank, (3) control system, (4) pressure regulators, (5) monitor valve/water meter, (6) head lines or main lines for conveyance of water from pump set to the field where water is to be delivered, (7) laterals (pipe) to carry water to plant rows and (8) the emitters/dripper through which water is finally released at a distance of 5 to 25 cm from the plant base. A schematic diagram of the drip system is shown in Fig.

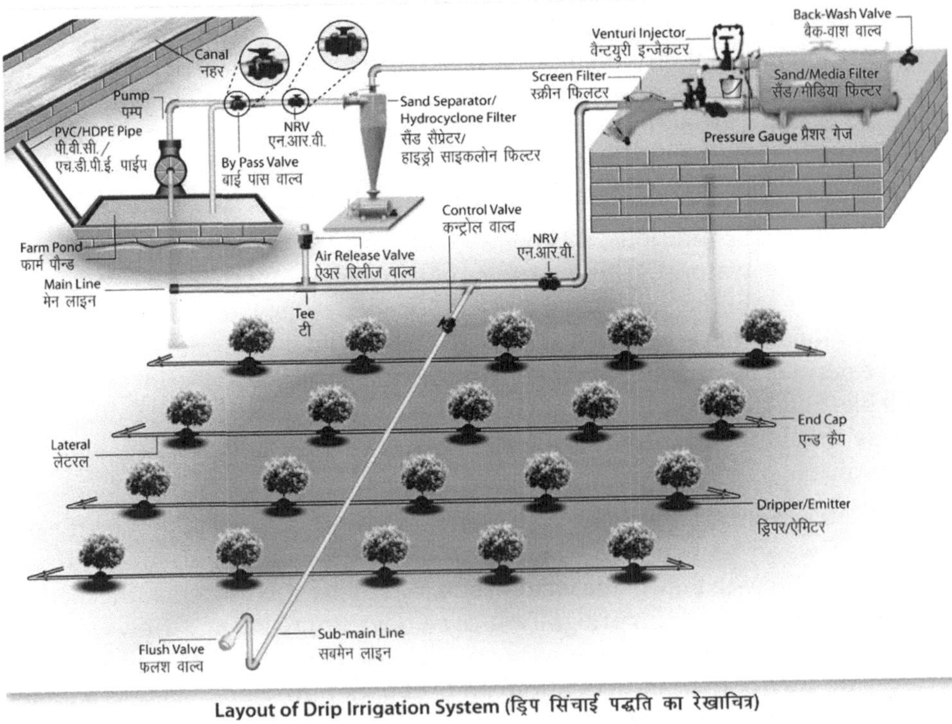

Layout of Drip Irrigation System (ड्रिप सिंचाई पद्धति का रेखाचित्र)

The water source may be well, canal, reservoir, river or tank but the important aspect is adequate quantity and quality of water. The pump set must be capable of supplying sufficient pressure and quantity of water to the system. A good filter is essential to keep away foreign material such as silt, sand and algae.

It is desirable to have a nutrient tank so that nutrients can be applied along with water rather than mechanical application. The control system, an important component, automatically monitors and regulates the pressure in the lines to keep the system at the optimum pressure required for uniform water distribution. The header lines carry the water to the lateral lines. The emitters are the flow regulators which are placed on the lateral lines.

Types of Drip Irrigation

Depending on the operating pressure at the upstream end, there are two types of drip irrigation.

1. **High pressure drip system:** In this system the operating pressure is 30 psi or above.
2. **Low pressure drip system:** In this system the operating pressure is less than 30 psi. This system suits to Indian conditions more due to economic considerations being cheaper.

Advantages of Drip Irrigation System

Drip irrigation system offers a number of advantages over other systems of irrigation like:

1. **Water Saving:** In this system water is applied directly to the root zone, eliminating wastage. The water saving may range from 30 to 70 per cent depending upon the area covered by the plants. This saving is achieved because of : less surface soil is wetted, less water is lost through wind evaporation unlike sprinkler; only required soil is wetted; water is evenly distributed throughout the root zone; no percolation loss; no water is utilized by weeds.

Water Loss under Various Irrigation Methods

Methods	*Normal*	*Hot Climate*
Surface irrigation	30 – 50 per cent	35 – 50 per cent
Sprinkler irrigation	6 – 9 per cent	10 – 20 per cent
Drip irrigation	1 – 2 per cent	2 – 3 per cent

2. **Labour saving:** This system eliminates the need for constructing borders, bunds, and labour intensive works associated with conventional irrigation techniques, channel or sprinkler irrigation, thereby cause 60 to 90 per cent labour saving. Labour saving is due to elimination of scrapping and tilling due to constant optimal oxygen/moisture relation, lack of conic

between irrigation, spraying and harvesting; the soil needs no furrow renewals, nor it is eroded and weeding is eliminated.

3. **Use of lower quality water:** In this system water is applied continuously and the root zone is kept wet constantly, therefore, salt concentration moves away from the root zone and therefore, permits use of low quality water as compared to other systems of irrigation. Drip irrigation takes highest salt concentration away from the plant root zone. In addition, since a much smaller quantity of water is applied to the soil, the total salt thereby applied is likewise lower.
4. **Increase yield and plant vigour:** Based on the individual plant's requirements, young trees develop significantly faster due to constant application of water to the soil without disturbing the soil structure. This maintains soil moisture at optimum level eliminating water stress resulting in greater vigour, better establishment and high productivity.
5. **Non-interference with cultural practices:** Most cultural practices may be carried out during the drip irrigation because the area immediately around the plant is wetted. Inter cultivation is easy in orchard crops.
6. **Use of hilly terrain and problem soils:** Drip irrigation permits the utilization of land in hilly areas which is difficult to be leveled economically. It also permits use of soils with physical problems like slow percolation rate, low depth *etc.*
7. **Reduced weed growth:** Since water is applied to the restricted area, widespread weed growth is inhibited due to restricted water supply.
8. **Saving of nutrients:** In this system nutrients are directly applied to the root zone along with water, therefore, chances of leaching losses are minimized resulting into saving to the tune of 30 to 60 per cent. In peach there is record that with 40 per cent of fertilizer application production and quality are not lowered. Some of the other advantages are:
   - **a.** It is not susceptible to wind
   - **b.** It works at a very low pressure
   - **c.** It can irrigate at any hour of the day
   - **d.** It causes no erosion even at steep slopes and high irrigation rate
   - **e.** Land levelling is not essential
   - **f.** It helps retain the soil structure for better respiration
   - **g.** Increased water holding capacity
   - **h.** It is ideal for poor soils
   - **i.** It is suitable for green house and made up soils
   - **j.** It causes no pollution due to leaching of nutrients
   - **k.** Less herbicide requirement
   - **l.** Less diseases-since stem and leaves are not wetted

Disadvantage of Drip Irrigation System

a. Higher initial investment.

b. Clogging of drippers due to oxidants, dioxides and algae.

Looking to the advantages that this system offers, it is an ideal system for arid and semi-arid regions for fruit crops, where water is scarce, soils are light and evaporative demand is higher.

Requirements for Drip Irrigation

1. **Choice of the crop:** It can be used for crops grown in rows with well marked distance between plants in the row. Thus it is highly suitable for orchard, plantation and vegetable crops and the crops with high value.
2. **Field preparations:** Field preparation is significant in laying out of drip system as undulations in the field have great impact on the energy gradient of flowing water through pipes in a low pressure drip irrigation system. Therefore, before laying down of mains and sub mains field should be properly prepared and uniformly leveled. Laterals should be laid on zero slope or down the slope to add to energy and hence in the uniformity of water distribution. Laterals should not preferably be laid down along the upward slope of more than 1 per cent or the pressure at head will have to be increased considerably.
3. **Planting technique:** A technique adopted should be such that it may reduce the number of laterals without affecting planting density and production but definitely reduces the cost of system. For fruit crops rectangular system or square system will suit.
4. **Irrigation scheduling and frequency:** The basic objective of the drip system of irrigation is to provide each plant with continuous supply of water which is just sufficient to meet the evapo-transpirational demand. Thus it is desirable to irrigate daily at the end of day break. To calculate daily consumptive use of water, potential evapo-transpiration values could be multiplied by crop factor approximating 0.7. This approximate quantity of irrigation requirement may be translated into period (in minutes) for which the system may be run in usual could sunny or cloudy day. To prolong the interval of irrigation, the water supply could be increased beyond consumptive use. These could be standardized for each crop through experimentation.

Above all it can be concluded that drip irrigation system is highly economical and useful for orchard crops in arid and semi-arid conditions. It also provides scope of using saline water in light soils. It has got only one limitation that it calls for higher initial investment at the outset.

Chapter 5

Soil Moisture Conservation

Soil and water are important components in success cultivation of crops. They assume special significance in arid conditions where soil is generally of poor class and moisture is limiting. Therefore, soil and soil moisture conservation assume special significance in these areas. Soil conservation would refer to maintenance of soil nutrients, soil structure, soil micro organisms, soil organic matter and the physical existence of so while soil moisture conservation would refer to saving of water losses from runoff, leaching and evaporation and making available for crop production. In general context, a number of soil and soil water conservation methods have been suggested in various fruit culture books but in reference to arid conditions, it is necessary (a) to prevent soil erosion, (b) to increase in soil moisture conservation, and (c) to remove runoff water out or the fields without significant erosion hazards. Therefore, techniques of conservation relevant only to arid region shall be discussed.

Techniques of Soil Conservation

Agriculture is based on the layer of top soil which, on an average, is only 7 inch. It is estimated that the formation of an inch of top soil takes about 300 to 1000 years, whereas, its depletion can be rapid. Therefore, it needs to be conserved or renewed. This loss of soil is mainly due to water and wind can be remedied. Soil conservation, therefore, aims at effective control of soil erosion. The rate of soil erosion depends on rain's erosiveness, and soils erodibility, the length and steepness of slopes, the cultural practices used, the stage of crop growth and supporting conservation practices applied to the land. In mathematical terms-Erosion is a function of Erosivity and the Erodibility which is under the influence of many factors.

$$\text{Erosion} = f\,(\text{Erosivity}) \times \text{Erodibility}$$

Therefore, soil conservation would call for proper utilization of land *e.g.* steeper slopes should be used for forest or permanent pasture; gentler slopes may be used for fruit trees and other crops that provide some protection to the soil. There are different methods

1. Maintenance of Soil Cover

In this technique permanent cover of vegetation is to be provided. This is considered to be the best technique of soil conservation in arid, as it would check erosion both by wind as well as water, which are of common occurrence in such areas. This will also permit slow seepage of water and improve the physical condition of the soil. Tree crops reduce the wind speed, and help stabilizing the soil by their extensive root system. Requirement of tillage, which accentuates 'the problem considerably, is also minimum for tree crops. This suggests plantation of tree crops in arid region for soil and water conservation. In an orchard, in arid, techniques used for increasing soil cover include, sod culture, intercropping, cover cropping and mulching.

2. Speed Barriers

The speed with which erosion takes place is determined by intensity and amount of rainfall and speed with which it is flowing. Any technique which hinders the speed and allows the seepage of water in soil shall reduce the erosion like deep ploughing, by increasing organic matter in the soil and by increasing drainage. Speed can be reduced by contour tillage ploughing in the direction of contour rather than along the slope.

3. Use of Inter Tillage or Strip Cropping

This technique calls for alternate rows a of row crops on the contour are effective for holding soils on hilly or rolling lands. The soil related crops could be biennial or perennial grasses or legumes which can also provide source of mulch for other soils and crops.

4. Terracing

Technique is highly useful in steep and rolling land where earth terraces are constructed on the contour. Water is held by the terraces until it penetrates the soil. Thus it serves twin purpose of conserving water as well as prevention of top soil loss. Fruit crops are easily adapted to terraces. Trees can be planted on the parapet of the terrace or between terraces. Inter space can be covered with grass to provide ground cover and organic matter. The most common type of terrace on gently sloping land is the broad base. The steep back slope terrace is most common on steeper land.

5. Plantation of Windbreaks

To avoid wind erosion, barriers are planted in terms of windbreaks right angle to the wind direction which fast growing, thick canopied plants are chosen. This aspect has been discussed in detail earlier.

Techniques of Soil Moisture Conservation

It is a fact that distribution of vegetation over the earth surface is controlled by the availability of water than by any other single factor. In arid zone, we are aware that, rain is exceedingly low and its distribution is erratic leading to low plant stand and productivity. To make it still worse, the water storage capacity of soils in these areas is very low, being mainly poor textured and shallow soils, and is coupled with high evaporative losses. To make fruit crop production a success appropriate techniques will have to be adopted to increase *in situ* water conservation by reducing runoff, leaching losses and evaporation. Some of the relevant techniques for arid region are:

a. Water Harvesting

Water harvesting practice of runoff collection in ponds and recycling it to supplement plant requirement is not quite feasible for large area because highly porous soils and high cost involvement in construction in cementmortor. This is only feasible for commercial nurseries. Only feasible proposition is *in situ* water harvesting. However, success of this runoff farming is primarily a function of effective utilization of runoff by crop plants. For this either we select a crop in which water requirement synchronizes with runoff events or select a crop which can utilize water available in deeper strata of soil. For this drought hardy, perennial crops with deep root system have preference over all other crops. In this case, the idea is that for each tree micro-catchments can be developed and orchards be raised. The size of the micro-catchment will depend upon local annual rainfall, catchment characteristics, and the water requirement and rooting characteristics of the tree species.

The advantages of micro-catchments in runoff collection and possibility of profitable orcharding. They reported that under 100 mm rainfall conditions in the Adavat highlands of Israel, a 0.1 ha catchment with 10 per cent slope would produce 160 m^2 per ha of runoff compared to only 50 m per ha from a 20 ha catchment. According to them further sub-division of catchment to micro- catchment considerably increased the runoff.

For planting fruit trees in Israel, they made square basin, shaped in the form of a flat inverted truncated pyramid at the lowest point of the Negarin (micro-catchment). The fruit trees were planted on 125 to 1000 m catchments. Among the various fruit crops tried at Shivata, fig, vine, pomegranate, and carobs survived the best of which carobs were most drought tolerant. At Adavat, peach, apricot, almond and longan berry were the most, successful. For rocky and gravelly habitats o Indian arid regions, the estimated catchment size for various fruit trees.

Similarly arid sandy soils, where runoff is about 25 per cent in 250 to 300 mm annual rainfall which may be more than 40 per cent in flood years with 400 mm rainfall, water availability to the tree, can be increased day runoff concentration from an appropriate size of micro-catchments to optimize production. The catchment area can be all around the tree or on its two sides depending on topography. An estimate of the catchment area per tree and plant population per ha for some arid fruit trees.

Thus by adopting appropriate water harvest technique crops could be raised profitably and water saved from loss in runoff.

b) Increase in Moisture Storage Capacity of Soil

To avoid water loss through leaching, which is a common problem in light textured soils, their water holding capacity can be increased by (i) increasing the content of soils, fine fraction clay, silt and organic matter and (ii) placement of slowly permeable layer at the bottom of the profile/or appropriate depth to arrest deep percolation. In other words, provision of subsurface barrier for which Bentonite clay barrier has been found to be highly suitable which has very low hydraulic conductivity, 10 to 15 T Bentonite per ha is considered to be appropriate.

For increasing water holding capacity of the soil application of pond silt, exfoliated vermiculite and FYM are recommended,

c) Crescent Bunding and Opening of Catch Pits

Crescent shaped or semi-circular bunds with a diameter of 6 to 13 meters (depending on the space requirement of the trees) are prepared and catch pits are also dug at the same time on the upper side of the slope. The trees are planted in the centre of the crescent. While crescent bunds help collect rain water, the catch pits conserve the same. New catch pits are opened as the old ones get filled up with the silt or organic matter.

Similarly planting on terraces and in trenches also helps in conserving moisture and sustained fruit production. To avoid water losses through evaporation use of windbreaks, mulching, and application of soil amendments would help.

Moisture being the most important input it needs to be conserved efficiently to make fruit cultivation possible in arid conditions and depending on the situation appropriate techniques should be adopted.

Chapter 6

Soil Plant Water Relationship

Soil plant water relationship related to the properties of soil and plant that affect the movement, retention and use of water. Soil provides the room for water to be use by plant through the root present in the same medium. Water as such and also as a carrier of large amount of nutrient is require in large scale for successful growth of the crop. Water is present throughout the plant body. From soil water around the root the liquid vapour inter phase in the leaf. The evaporative surface of the leaf mesophylls cell discontinuity between the water with in the plant body and water vapour in the atmosphere. Every individual growing cells surrounded by and impregnated with water. Water is most abundantly present in actively growing plant cell. The availability of soil water, the plant root demand and water vapour are the most important ecological factor governing the distribution of higher plant species on the surface of earth. In most cell/tissue of the higher plant water constitute more than 80 per cent of fresh weight. The water content of some growing cell may rise to 90 per cent or more but in dormant seed and bud water content may be 10 per cent or less.

Function of Water in Plant

The significant of water to the life of higher plant can be emphasized by highlighting its function which is as under:

1. Water is a constituent of protoplasm.
2. Water is a solvent of mineral nutrient through which they find in tree in the plant body and as a carrier of nutrient with in plant from one organ to another.
3. Water is a medium in which many metabolic reactions occur.

4. Water is reactant in a number of metabolic reactions *i.e.*, certain reaction in TCA cycle.
5. In photosynthesis the hydrogen atom in water molecule is carbureted in to organic compound and oxygen atom is released at O_2.
6. Water is important for the turgidity of the growing cell. Thus it should maintain their form and structure.
7. Gain or loss of water form cell and tissue is responsible for a variety of movement of the plant part like opening of stomata, nocturnal folding of leaves of the some plant, opening and closing of the flowers in certain plants and sensitivity to touch of leaflet as in *memosa pudica*.
8. The cell elongation phase in growth is depended on water.
9. Water is a metabolic end product of respiration.
10. More water is absorbed by plant and greater amount of water or loss by plant than any other substances.

Since which water to the plant is important as a constituents, solvent, reagent and maintainer of cell turgidity therefore knowledge of soil plant water relation is absolutely necessary for successful and economic crop cultivation under dryland condition where water is a scarce input and soil are of poor quality. And efficient utilization and management of available water is therefore essential in an assured crop production program particularly in dryland area. Where crop production suffer due to scarcity and irregular distribution of rainfall.

Water Potential

The behavior of water and its movement in soil and plant is based on potential energy relationship. Water has capacity to move from one area of high potential energy to low potential energy. The potential energy in an aqueous system in expressed by comparing it which the potential energy of pure water. Water potential is diagnostic tools that enable to assign prestige value to the water status in plant cell and tissue. The lower water potential plant cell or tissue greater is its ability to absorb water. Conversely the higher water potential the greater is the ability of tissue supply water the other more desiccated tissue and cell. Thus water potential is use to measure water deficit and water stress in plant cell and tissue. Since the water in plant and soil is usually not chemically pure due to salute and physically constrained by the force such as polar attraction gravity and pressure. Potential energy of water is called the water potential.

The greek later psi (Ψ) designated the water potential in a system weather a plant cell or tissue are a soil sample in which plants are grown. Water potential is expressed either in bars or mega Pascal (1 mpa = 10 bar). The pure has a water potential equal to zero bar. The water potential in plant tissue and soil is less than 0 bar and hence a negative number.

In general leaf of the most plant rooted in well water soil or likely to have water potential between about -2 and -8 bar. Decrease water supply leaf water potential become more negative than -8 bar and leaf growth rate will decline most plant tissue will cease growth completely when water potential drop to about -15 bar.

Soil Water Availability

The plant root growth in moist soil and extract water until the critical water potential in the soil is reached that root are no more capable of drawing water. The water that can be extracted from soil by plant roots termed as available water. This is the difference between the amount of water in the soil at the field capacity and the amount of water in the soil at permanent wilting point.

The availability of soil moisture is affected by the collindal property *i.e.* the surface area of the soil particle *i.e.*, soil texture a clay loam soil hold about 20 per cent of its weight as available moisture where as the courser texture soil such as fine sand, only hold about 7 per cent. On a soil volume basis the clay loam at field capacity will hold about 17 cm available water per meter of soil depth where as the fine sand will hold less than 8 cm. the arid soil of India also vary considerably in their characteristic. Therefore their moisture retention capacity also varies. The capacity of different soil available water is as fallows.

Soil Types	*Water Content as Percentage of Dry Weight at*		
	0.3 bar	*15 bar*	*Available Moisture (per cent)*
Sandy loam	12	05	07
Loam	20	08	12
Silt loam	28	12	16
Silty clay	41	22	19
Clay	45	26	09

Soil Water Movement and Uptake by Plant

Soil water is confined between soil liquid and liquid air inter phases. It may occur a thin film on the solid or as pore water when in excess. Its movement within soil and uptake by plant can be divided into following step:

1. **Movement of water through soil**: The movement of water through soil depends on differences in water potential of soil and water potential of roots. If the water potential of soil is greater than water potential of root the movement occur.
2. **Movement of water a cross the roots**: The roots are major site of water intake the zone of maximum water absorption lies 20-200 mm form the root tip behind the meristem which is characterized by the present of root hair which will increase the area of absorption. There are three pathways for movement of water through roots.

a. Through the cell wall.

b. Through cytoplasm via. plasmodesmata.

c. Through cytoplasm via. Vascular spaces of cortex.

3. **Movement of water through vascular system:** Xylem is the vessels through it water inter from root to leaves. The flow of water through xylem is by bulk flow caused by difference in pressure potential. Negative pressure is transmitted to liquid continually within xylem and root sap. The breakage of this column is prevented cohesive phase between water molecule and adhesive force operative between cell wall and water.

Ascent of Sap

Ascent of sap refers to up word movement of sap which takes place through xylem stream. The several theories has been proposed for movement of sap.

1. Vital theory
2. Root pressure theory
3. Physical theory
4. Atmospheric pressure
5. Imbibitions
6. Capillarity
7. Cohesion theory

Water Movement to Leaves

Through vascular tissue water reaches leave. Water move to leaf parenchyma from where evaporation takes place in possibly these two paths.

1. Movement of water through leaf parenchyma to liquid: Air inter phase with the leaf and then vapour phase.
2. The movement of water through leaf parenchyma to the liquid: Air inter phase in the outer surface of the epidermis cell and then to the vapour phase through the cuticle to the air.

Water Uptake and Movement in Plant

The complete path of water from the soil to the plant to the atmosphere forms a continuous system. Which may be analyze by evaluating the potential difference between soil and atmosphere in connect with root and leaf respectively. The path of water may be divided into four sequential process as follows:

1. The supply of liquid water to the root surface.
2. The entire of water into the roots.
3. The passage of water in the conducting elements.
4. The movement of water vapour throughout the leaf.

The rate of water movement in everywhere proportional to the potential gradient and inversely proportional to the resistant to flow.

Air usually has an extremely low water potential compare to plant and soil. Since a living leaves usually has a water potential greater than -15 bar. There is a steep energy gradient and continues movement of water as vapour loss from leaf to air. At night when there is no loss of water as vapour loss from leaf to the air. The water potential of plant would come close to equilibrium with water potential of soil. When the stomata are open the water loss from leaf is continuous. Which lower water potential of leaf below that water potential of petiole. Since water flow from higher water potential to low water potential the water will flow from petiole to leaf. The water flux reduces the water potential of petiole which was as equilibrium with the water potential of stem, so water move from stem to petiole. This energy gradient is continuous down to the roots and soil. In other word the system develops a water potential gradient from soil to air.

To elucidate the point as are clearly when we irrigate soil and allow loss of gravitational water. Water potential of soil is -3 bar at night the stomata are closed so water move into plant and creates water potential of soil and plant equilibrium. During day time stomata are open and transpiration occurs. As the water is loss form leaf. Water potential of leaf become reduces, initiating the water potential gradient. This provide energy differential to cause water movement from the soil to replace the water loss by transpiration. At night stomata close and transpiration is reducing to near zero. However water will continue to flow in the system until the water potential of plant and water potential of soil again approach equilibrium. As water is removing from soil by the plant, water potential. As soil become lower and water potential of leaves become relatively lower, establishing a water potential gradient for continuous uptake. The transpiring plant always show a negative potential as the water is loss from the soil water potential of leaves root and soil go below the water potential of -15 bar. This usually indicates that the water availability to the plant is not enough to prevent wilting and recovery will not occurs unless water is added to the soil. The approximate magnitude of water potential in the soil plant atmospheric system is as fallows.

$$\text{Water potential} = \frac{\text{Potential gradient}}{\text{Resistance}}$$

Components	***Water Potential (MPa)***
Water potential of Soil	0.1
Water potential of Root	1.00
Water potential of Stem	1.5
Water potential of Leaves	1.5
Water potential of Air	100

Chapter 7

Water Stress and Plant Growth

The term water stress refers to the situation where cell and tissue are less then truly turgid. It occurs whenever the loss of water in transpiration exceeds the rate of absorption. It is characterized by decrease in water content, osmotic potential, total water potential, accompanied by loss of turgidity, closure of stomata, decrease in growth *etc.* its sever water stress result in drastic reduction in photosynthesis and disturbances in many other process, cessation of growth and finally death. Impact water stress practically affects every aspect of plant growth modifying the anatomy, morphology, physiology and biochemistry.

The affect of stress during the vegetative growth and development of smaller leaf which can reduce leaf area index at maturity and result in less light intersection by the crop. Chlorophyll synthesis is inhibited at greater water deficit. One of the consequences of water stress in plant is a change in structure and configuration in enzyme as a result enzyme activity change and metabolic rate of affected adversely. The amino acid and protein metabolism of the plant can be affected by water stress condition.

With reduce water potential plant hormone also change into concentration specially abscise acid increase in leaf and fruit. Abscise acid accumulation induces stomatal closure, which result in reduce carbon dioxide assimilation, older leaf and fruit often abscise. If accumulation is high. Cytokinin and ethylene can counter abscise acid. Effect of often increase when ABA in increase. This may account for more rapid fruit ripening under water stress. Under moderate to severe stress condition. Their water significant increase in level in free arginine and especially proline. Proline seems to aid in drought tolerance acting as storage pool for nitrogen or as a solute molecule reducing the water potential of cytoplasm.

The stream of level of stress (water potential greater than -15 bar) respiration, Carbon dioxide assimilation, assimilate translocation and xylem transports rapidly diminish to lower level while the hydrolytic enzyme increase. The major effect of water stress on tree crop is as follows:

1. Once the actual wilting occur inhibition of growth coupled with other phenomenon like enhancement of RNA destruction which decline the protein.
2. The young tree growth and bigger determine the precocity of bearing and size which are delay and reduce respectively.
3. In older tree growth and vigour, yield and productively goes down.
4. Water stress reduces assimilation by reducing photosynthesis and increasing respiration and reduces absorption of nutrient.
5. Root growth is adversely affected beyond the point of repair.
6. Quality of fruit is reduces like Shrinkage of citrus fruit and small size.
7. Water stress promotes more root than shoot formation. A high root shoot ratio reduce crop yield as compare to adequate moisture develop lower root/shoot ratio.

Chapter 8

Evapo-Transpiration

Evaporation: Evaporation is the process during which liquid change in a gas. This is the only form of moisture transfer from land and ocean in to the atmosphere or the loss of moisture from soil, land, ocean *etc.* is known as evaporation.

Transpiration: Transpiration is the process by vapour pressure leaves the living plant body and inter the atmosphere or the loss of moisture from plant is known is transpiration.

Evapo-transpiration or consumptive use of water: it denotes the quantity of water transpire by plant during their growth in the plant tissue plus moisture evaporation from the surface of soils and vegetation

The term consumptive use is the used to designate the losses due to evapo-transpiration and water that is used by plant for it metabolic activity. Since the water is used in actual metabolic process is significant (less than 1 per cent of ET) the term consumptive use is generally taken equivalent to ET. There is most important is working out the water requirement of plant/crop.

A. Environmental Factor

The environmental or evapo-transpiration is called atmospheric demand or evapouratory demand. The greater the atmospheric demand. The faster would be water loss from a free water surface. The fallowing factor affects the atmospheric demand.

1. **Solar Radiation:** Solar radiation absorb by leaf 1-5 per cent use for photosynthesis 75-85 per cent is use to heat the leave and for transpiration. Therefore increase the solar radiation will increase the atmospheric demand of water.

2. **Temperature:** Increase the temperature the Capacity of air to hold water which means the greater evaporation demand.
3. **Relative humidity:** Higher level of humidity in the air the water potential of air would be higher amounting to lesser atmospheric demand and vice versa.
4. **Wind:** Transpiration occurs when water defuses through the stomata high velocity and hot desiccation wind increase evapo-transpiration.

B. Plant Factor

1. **Stomata closure:** The most water loss from plants occurs through stomata because of relative impermeability of cuticle due to cutin and waxy layer. The little transpiration occurs when stomata are close but stomata open wider more water is loss. Stomatal opening again under the influence of many factors which light and moisture level are prominent. Low moisture and low light caused stomatal closure.
2. **Stomata number and size:** The productive crop plant leaves have numerous stomata of which major concentration is on lower surface. They would in general have less effect than the stomatal opening and closing on water loss from plant body.
3. **Leaf amount:** More leaf surface area the greater the evapo-transpiration.
4. **Leaf rolling and folding:** Much water has mechanism in leaf that favours reduce transpiration when water becoming a limiting by rolling as in case of maize and folding in soya been
5. **Root depth and proliferation:** The availability of extraction of soil moisture by the crop highly dependent of rooting depth and proliferation. Deeper rooting increase water availability and proliferation Increase water extraction from until volume of soil.

Importance of Transpiration in Plant

Transpiration is loss of water from the plant evaporation modified by plant structure and stomatal behavior. This loss of water in one way is by product of carbon fixation. Carbon dioxide which provide the carbon necessary for growth of internal tissue the water film surrounding the spongy mesophyll of leaf as this film evapourate it is replenished from the tissue of which enter the drought water from vascular tissue. Therefore this transpiration loss of water by plant can be considered as an exchange of water for carbon. It also benefit the plant by cooling effect, help in growth and development and mineral uptake.

1. **Cooling effect:** The cooling effect of transpiration help to keep the temperature of leaf below the lethal level during the period of abundant solar radiation. Stoppage of transpiration may produce 8-10 °C rises in leaf temperature which may prove lethal to the plant. In arid region where

soil moisture is less than optimal stomata vary often close during hottest hour of the day when cooling effect would most advantageous.

2. **Effect of growth and development:** The bud of hardy pear cease to grow under condition of high humidity since transpiration is negligible under the condition of high humidity. Transpiration is necessary factor in normal growth of the plant when transpiration exceed that of absorption water deficit may occur and wilting may take place. This is the detrimental to the plant and if carried to extreme may result in death of the plant.
3. **Effect of mineral salt absorption:** Salt absorption is predominately active process (requiring metabolic energy) and that only a small amount of salt is absorb positively a result of water uptake. Once the absorb salt has been dump in to xylem duct of root. Transpiration definitely influences their translocation and distribution in the plant. In this respect the transpiration extreme provide and efficient means of transport and distribution of salt have been absorb by root.

Control of Loss of Water through Transpiration and Evaporation

Losses in fruit yield from moisture stress exceed those from other causes including paste and diseases. This can partially be reduced by reducing the rate of transpiration from plant and evaporation from soil surface. There for reduction of transpiration and evaporation can be important both for decreasing irrigation water requirement and under arid condition for alleviating the water stress. There are several possible approaches to reducing the rate of transpiration in plant.

A. Control of Loss of Water through Transpiration

1. **Choice of the plant:** Mesophytic plants growing under dry condition transpire more than 99 per cent of all water they absorb. This is necessary for photosynthesis under identical condition. Plants vary in their capacity to transpire water. This variability depend upon many plant factor like:
   - **a.** Efficiency of root system
   - **b.** Leaf area
   - **c.** Leaf arrangement and structure.
   - **d.** Stomatal behavior.

   Most of the tree crop have stomata located only on lower surface of the leaf compare to both surface in annuals likewise there are many crop shed off leaves during stress period and reduce transpiring process to overcome stress effect like Ber, Custard apple, Pomegranate, Gondy, *etc.* likewise plant also vary in their water use efficiency. Pineapple is shed to produce as much dry matter per year as sugarcane but use only 10-12 per cent. The low water loss occurs because the stomata of pineapple are closed most of the day. Pineapple as "crassulacean acid metabolism" which enable it to store CO_2 as organic acid at night and convert it into carbohydrate during day resulting in efficient production of dry matter. Therefore by making suitable choice of crop stress could be overcome.

2. **Modification of leaf reflection:** In this approach it intended to regulate gas exchange of leaf by modifying their energy balance. Energy to plant during day time comes from solar radiation at super optimal intensity (1.5 cal per cm^2 per minute) as photosynthesis saturate at (0.2-0.5 cal per cm^2 per minute) and 60 per cent of solar radiation of wave length more (0.7 μ) is not use in photosynthesis. This energy input can be reduced by increasing plant reflectivity by using effective zink oxide, Kaolinite, chalk *etc.* alone or in combination with other anti transparent. These chemicals have been use and known to reduce temperature of the plant part and vapour pressure of leaf resulting into transpiration without affecting photosynthesis while selecting such material it should be seen that the material should not be toxic and spread evenly of plant surface and must be stable and permeable to gases to be non interfering with photosynthesis and transpiration.
3. **Increase in leaf resistance:** It refer to coating of leaf with inert material to increase leaf resistance to water loss. For these two types of compound are uses.
    i. Film farming compound:
    ii. Compound affecting opening of stomata

    Film Farming Compound: A number of chemicals have been use by various purpose a material with high permeability to carbon dioxide than water vapour is require. The following commercial product has been better.
    a. Wilt proof
    b. Clear spray
    c. Mobi leaf
    d. Folicote
    e. Vapour gard

    Silicon emerson or plastic film and low viscosity waxes. The ideal anti transparent should fulfill the following criteria
    1. They should be non toxic to human and plant.
    2. They should have low mobility when spread on plant to be influencing only gourd cell and cell immediately around them with no phytotoxicity.
    3. They should cause no permanent damage to the stomatal mechanism or to the structure of gourd cell.
    4. They should be persistent for several weeks.
    5. They should be cast effective.

Compound Affecting Opening of Stomata: The certain insecticide. Fungicide, herbicide and metabolic inhibitor or know to reduce rate of transpiration through partial stomatal closure. Some of these are phenyl mercuric acitate (PMA) Decenyl succinic acid (DSA), Atrazine, cymezine, Alkenyl succinic acid, alpha-hydroxysulphonate, Beta-hydroxy quinol, Sodium azide, Abscisic acid (ABA), acetyl alcohol *etc.*

By selecting these chemical the following point should require configuration:

- **i.** Low mobility within the tissue.
- **ii.** It should confine to leaf epidermis to avoid toxic side effect.
- **iii.** Persistency
- **iv.** Photosynthesis activity should not be hampered.
- **v.** Non expensive.

Among the chemical mentioned above PMA is most commonly used at foliar spray at a concentration of about0 (10^{-4} M). It influences stomatal opening by influencing protein component of gourd cell, membrane and potassium transport and through blockage of photosynthesis, and thus by accumulation of respiratory carbon dioxide or by removing energy supply of potassium uptake.

ABA is an indigenous inhibitor known to regulate stomatal opening through changing the osmotic potential of gourd cell by inhibiting potassium K^+ influx, starch hydrolysis and cyclic photo phospho relation. It will not have any adverse affect of photosynthesis.

Carbon dioxide is another substance *i.e.* very effective anti transparent raising the concentration of CO_2 only slightly in the air surrounding a leaf from the natural 0.03 - 0.05 per cent is known to induce partial closure of stomata

The growth retardant ALAR and Paclobutrazol may reduce water use by decreasing leaf area of the fruit crop.

4. **Erection of wind break:** Wind break is a narrow strip of tree planted around the farm, garden and orchard to have protective barrier against hot and cold wind current. Wind breaks help to deflect and filter through the wild current. Their by reducing the velocity of wind resulting losses displacement of wind around tree and cause reduction in transpiration and evaporation. The tree suitable for windbreak in dryland region should have following properties.

    - i. The species should be hardy and fast growing.
    - ii. Should be wind firm.
    - iii. Should have dance canopy.
    - iv. Should have dense foliage with erect and drooping branches.

v. Should have long life.

vi. Should have deep tap root system with minimum lateral spread.

vii. Preferably a leguminous tree with canopy to farm better under need to add organic matter to the soil.

The site reducing water loss through transpiration and evaporation they have following other beneficial fruit cultivation effect.

a. They protect fruit plant and cold and scorching wind thus reduces the losses.

b. They prevent flower and fruit abscission by reducing wind velocity and moisture losses.

c. They prevent breakage and uprooting of fruit tree.

d. Due to increase infiltration and reduce water loss it increases availability of water to fruit tree.

The important windbreak trees are *Acacia tartalis, Cassia siamea* and *Prosopis julifera* are some useful tree species suitable for windbreak in arid condition.

5. **Training and pruning:** Summer pruning can be use to control orchard to water use and sever pruning have been found to reduce transpiration and survive better under drought condition.
6. **Crop load:** Reducing fruit load is also practice to improve water balance having fruit cropping tree put less carbohydrate in to vegetative and root growth thus reducing water uptake capacity.

Control of Loss of Water Through Evaporation

Loss of water through evaporation can be minimizing by several practice.

1. **Tillage:** Tillage is practices which loose the soil, kill the weeds and increase soil porosity. These entire factors affect soil water relation. It helps in this connecting capillary connecting from subsoil and farm soil mulch and help in conservation of moisture to be used only by plant root. Tillage of soil at appropriate stage is highly significant for moisture conservation.
2. **Mulching:** It refers to covering of soil surface with organic and inorganic material to the soil for moisture conservation. It not only help in conserving the moisture by reducing water loss through evaporation keep the weed under control. Mulches reduce evaporation by cutting of the radiation falling on soil surface and thus delays drying and reduce soil thermal radiation in day time.

   It also prevents soil erosion, add organic matter to the soil and increase it infiltration capacity. Material such as hay, straw, dry leaf, cut grass, saw dust, peanut hulls, peat mass *etc.* can be use. In recent year the use of polythene has become very popular they are being used for the control of soil born diseases paste and nematode especially in nursery.

3. **Use of soil amendments:** By increasing water holding capacity through incorporation of soil amendment water loss through evaporation can also be minimized by pond sediments, FYM and Vermiculite have tremendous capacity to increase water holding capacity of the soil and can be use to control evaporation.

Chapter 9

Orchard Management in Dryland

In dryland area availability of water and quality of underground water are the most important limiting factor in cultivation of fruit crops. Similarly adverse weather, soil salinity, shallow, rocky and gravely soil are the other limitation. To remove or modified these limitation a number of suggestion have been made by various scientist. Some of the most important practices are as follows:

1. **For Water Use Efficiency:** Water use efficiency is defined as ratio of crop yield to water used by crop area or evapotranspiration of crop area.

   Water use efficiency = Crop yield/evapo-transpiration of crop area.

   Water use efficiency in dryland horticulture can be increase by following way.

   a. **Provision of micro catchment around the tree:** In dryland tracts average yearly precipitation is very low which make good cultivation difficult. Since arid soil is high infiltration rate as well as high runs off losses due to concentrative rain. The idea here is develop catchment area for each tree. The size of which is determined by slop of land, water requirement of the optimum production, runoff coefficient and canopy of fruit tree, the several study has been done in National Research Center for arid zone, Jodhpur. Shown that Ber tree is canopy area of 36 m^2 can be successfully grown if provided with a catchment area of 75 m with a slop of 5 per cent this means that one Ber tree should be planted in an area of about 110 m^2. Likewise the plant population of fruit tree should be grown with runoff from different micro catchment area estimated and tabulated as fallows

Catchment Area per Tree and Plant Population per Hectare with different Slopes of a Sandy Catchment (annual rainfall 360 mm).

Fruits	Canopy Area (m^2)	Run off Supplement (mm)	Catchment Slope (Per cent)	Catchment Area (m^2)	Area Tree (m^2)	Population/ ha
Ber	36	300	0.5	100	136	74
			5	75	111	90
			10	60	96	104
Pome-granate	20	800	0.5	148	168	60
			5	111	131	76
			10	89	109	92
Guava	56	800	0.5	415	471	21
			5	311	307	27
			10	249	305	33
Aonla	56	500	0.5	259	315	32
			5	194	250	40
			10	156	212	47
Sour lime	36	1000	0.5	333	369	27
			5	250	286	35
			10	200	236	42
Custard apple	20	300	0.5	56	76	132
			5	42	62	161
			10	33	53	189
Lasoda	36	200	0.5	67	103	97
			5	50	86	116
			10	40	76	132

It is assumed that the runoff from hilly area in Indian arid zone would be about 50 per cent the catchment size for various fruit tree at given below:

Catchment Area per Tree and Plant Population per Hectare for some Fruit Tree in Rocky and Gravelly Slope (Annual rain fall-360 mm)

Fruits	Canopy Area (m^2)	Catchment Area (m^2)	Population per ha
Ber	36	60	104
Pomegranate	20	90	91
Guava	56	250	33
Fig	56	250	33
Aonla	56	156	47
Custard apple	20	35	182
Sour lime	36	200	42
Lasoda	36	40	132

b. **Crescent bonding and opening of catch pith:** The crescent shape are semi circular bund with diameter of 06-13 m depending on the spacing requirement of individual fruit tree are prepared and catch pit are also dug at the same time on the upper side of the slop. The trees are planted in the centre of the crescent. While the crescent bunds collect rain water the catch pith conserve the same.

c. **Planting on terraces:** In hilly area depending on the spacing requirement and slop of land broad terraces are prepared with the slop toward the higher elevation and trees are planted. Such terraces help in conservation of moisture and support plant growth.

d. **Trench planting:** Deep trenches (0.5-01 m) are dug a cross the contour and fruit tree such as Ber, Aonla, Pineapple and Custard apple are planted. The trenches collect rains water along with silt and organic matter and thus promote the tree growth.

e. **Wind break and shelter belt:** There is a heavy damage to fruit crop in arid tract by frequent wind by way of water loss through evapo-transpiration and also through dust deposition. Mechanical damage to plant and soil erosion. A single row plantation of *Accassia tartalis* for arid tract in north India have been found useful.

f. **Mulching:** The material such as hay, straw, cut grass, dry leaf and polythene sheet can be use for mulching. A mulch beside conserving moisture also suppresses weed growth, prevent soil erosion and add organic matter to the soil.

g. **Use of anti-transpirants:** Anti-transpirants is the chemical which when spread on the plants form a film which increase the diffusion resistance of water from stomata and thus reduce transpiration losses of water. Several anti-transpirants have been used in fruit tree. Like Acropyl in grape wine, Polycot in banana, Kaolinite in several fruit plants.

h. **Use of drip irrigation:** Drip irrigation is a method of watering plant at rate equivalent to its consumptive use. So that would not suffer any moisture stress throughout their life cycle. It is not only the efficient mode of water use but also important in management of soil irrigated with saline irrigation water. This method can also be implied for using saline water for irrigation without allowing accumulation of salt in root zone. Although initial investment is quite high but in long run it is quite economical and efficient and boon for arid region for high value crop like fruit tree.

2. **Fruit crop for sallow soil:** Generally soil of arid region have low depth and hard pan in sub soil. So such fruit crop should be selected which thrive well in such situation. Fruit crops like pomegranate, custard apple grow well on the other hand crop root of which penetrate the hard pan should be selected like Ber, Kair and Lasoda.

3. **Salinity problem:** Wide occurrence of salinity and saline underground water are the common feature of arid region of India they affect plant by changing osmotic potential and specific ion effect. Such problem tackled by following approaches.

   a. **Use of salinity tolerance plant:** This is wide variation in salt tolerance fruit crop. If potential salinity problem is indicated suitable crop can be selected. Date palm, Ber, Aonla, Pomegranate, Fig, Guava, Kair, Pilu have great tolerance to such condition. However most of the crops are relatively sensitive during early stage of the growth. Based on salt tolerance limit of the fruit crops have been classified in to five groups.

      i. **High tolerance** (40-50 ESP and 12-15 EC): Date palm, Ber, Tamarind, Lasoda, Wood apple.

      ii. **Medium tolerance** (30-40 ESP and 09-12 EC)**:** Aonla, Phalsa, Jamun, Pomegranate, Mulberry, Karonda and Pineapple.

      iii. **Low tolerance** (20-30 ESP and 06-09 EC)**:** Guava, Bael and Grape.

      iv. **Susceptible** (20ESP and 06 EC)**:** Banana, Papaya, Pine apple and temperate fruits.

   b. **Planting technique:** Soil working technique should be such that the rain water should utilize to the maximum extent possible and the salt concentration in root zone is kept to the minimum level through leaching. Micro water harvesting catchment of 4-6 per cent slop on either side of the plant row provide 60-80 per cent additional runoff water is root zone and help in leaching off accumulated salt.

      Piths prepared for planting tree could either be located in better patches are the piths can be filled with imported soil and separated from original soil by providing a polythene lining in between so that young plant can establish well. There after the tolerant plant species may not be injured. The alkaline soil deep pith with the addition of zypsium and FYM 20 Kg give better result.

   c. **Irrigation management:** Scheduling irrigation with saline water 4-5 day prior to normal irrigation schedule and providing additional depth on irrigation water to meet the leaching requirement. Maintain adequate water supply to the plant and low salinity in soil. Use of drip method is still a better method to manage saline irrigation water and salinity problem.

   d. **Use of salt tolerance root stock:** Few of the fruit plant are susceptible to sodic/saline soil condition with used tolerant/resistance root stock. Their cultivation is possible to great extent. Root stock must have capacity to reduce absorption of harmful salt *i.e.* Na^{+}, Ca^{++}, and SO_4^{--} from the soil solution. The higher prolein content provide tolerant against moisture and slat stress therefore root stock should have high prolein content in the plant system so that these can tolerate harmful influence of the salt. Few of the tolerant root stock are as under-

Mango: Neleshwari Dwarf

Citrus: Cleopetra Mendrin, Rangpur Lime, Trayer orange, Jalandhari Khatti.

Grape: Dogridge, Saltcrick, 1613, 1616, Talki 5a.

e. **Selection and popularization of tolerance cultivar:** Attempt have also been initiated in required to genotypic variation to salt tolerance in some of the selected fruit plant and their relative tolerance cultivars are as follows:

Aonla: Chaikaiya, Kanchn, Fransis (Hathijhul)

Ber: Banarasi Karaka, Kaithli

Grape: Beauti Seedless

Guava: Saradar Guava (L-49)

f. **Selection of suitable cultivar for dryland:** Since water is a limiting factor in dryland selection of early cultivar provide distinct advantage over late cultivar in this region. Some of the promising cultivar which have been identifying from different parts of the country for different fruit crops are as fallows.

Ber: Gola, Seb, Banarasi Karaka and Himdia.

Aonla: Chaikaia, Kanchan (NA-4), Krishna (NA-5), NA-7

Bael: NB-1, NB-4, NB-5, NB-7 and NB-9.

Custard apple: Bala Nagar, British Queens, Mamooth Red, Seetaphal, Washingatan.

Pomegranate: Ganesh, Jalaur Seedless, Brassein Seedless.

Guava: Saradar Guava (Lucknow-49)

Phalsa: Sharbati

Fig: Pune Fig, Blackichia, Brown Turkey

Karonda: Pink and White fruited type

Chapter 10

Plant Propagation of Arid Fruit Crops

Plant propagation means increase in number or perpetuation of the species by reproduction. The success in plant propagation depends on a thorough knowledge of the growth of plants, choice of suitable method and technical skill. The propagation of plants may be done either by the gametes or from the somatic cell, tissues or organs. In other words, propagation may be done by sexual methods or by asexual methods. This reproduction is achieved through sexual and asexual methods. Sexual propagation in which multiplication of the plant by seed or meiosis. It is also known as seed propagation. Sexual propagation is the raising of plants by means of seeds which are formed due to the fusion of male (pollen) and female (egg) gametes within the ovule of a flower. A seed is a miniature packed plant ready for sowing with nourshing tissues and a protective cover. A seed consists of three basic parts namely (i) embryo, (ii) the food reserves (endosperm, cotyledon) and (iii) the seed coat (testa). The embryo is a minute plant, carries all the genetic information and is capable of giving rise to a new individual in a congenial condition. An embryo may also be developed from the somatic tissues in some species without meiosis which is known as apomixis. Asexual propagation which means propagation of plant by vegetative part like stem, root, leaf *etc.* It takes advantage of the precise duplication of genetic information inherent in mitosis and is possible because of totipotency. In addition to sexual and asexual propagation there are two methods which combine feature of both: apomixis and tissue culture.

Asexual propagation permits cloning of plants, meaning the resulting plants are genetically identical to the parent plant. The population of plants generated from a single individual which maintain the same genotype is known as clone. Since the product of asexual propagation is reproduction from the vegetative parts of the mother plant and involves mitotic cell division in which the chromosomes duplicate and divide to produce two nuclei that are genetically identical to original nucleus. A number of asexual methods of propagation are known of which most relevant method could be adopted depending on the species and results desired. Use of suckers, air-layering, stooling, cutting, grafting, budding are most common methods. However an appropriate technique of propagation for raising fruit crops in arid region is absolutely necessary. Growing of fruit tree under arid conditions is a real challenge for the horticulturist and farmers as well. The extreme climatic conditions, dependence on meagre rainfall and varies of nature in these areas make up the job much more difficult. The propagation practices of various fruit crops in arid region should be such that it can overcome the adverse natural condition so that establishment and growth of the plants can be proper. In fact the method should be such that it provides unhindered development of natural root system so that plants can draw moisture from lower strata of the soil and overcome the problem of water stress during lean period the moisture availability. Therefore for arid regions following two methods which considered most appropriate are being given details treatment: soft wood grafting and raising of budded plant being most relevant for arid and semi arid region.

A. Soft Wood Grafting

This method consists of detached grafting on new terminal growth seedlings grown in situ. This method was standardised for mango but can also be employed successfully for aonla, cashewnut, guava, jackfruit, phalsa, sapota, jamun, with 70 to 100 per cent success.

1. **Selection of place:** After having made the choice of the place for fruit tree plantation depending on the space requirement and layout method pit size 1mx1mx1m are dug in summer and left open to sun heat for couple of weeks. The pits are refilled to a depth of 30 cm with a mixture of top soil 60 kg FYM, 2.5 kg SSP just before the onset of monsoon. After first shower of rains, the remaining top soil of the pit is refilled with a 10 cm layer of dry leaves over which a layer of 8 cm soil is made. Then three seeds are placed in the pit and duly covered with soil. These seeds will germinate with rain water. In case the rain are deficient the pits can be spot irrigated. Three seeds are placed in each pit to ensure some germination even if some fail and a layer of dry leaves in the middle of the pit is made to break the capillary movement of water from lower layers to top under force of evaporation. However this water shall be available to roots which are likely to penetrate down below. This in fact is a water conservation technique.

   A basin is made around the planted seeds to conserved rain water. After the monsoons have withdrawn a dry mulch is provided around the plants in basin to conserve moisture and to avoid excessive rise of soil temperature around the young plant being sensitive. These saplings can be used for grafting when they had grown for one year.

2. **The stage and site of grafting**: New shoots on stock plants of one or more years old emerge from March onward. These new shoots can be identified their tenderness and leaf colour which for example in mango shall be bronze coloured. It is only this tender growth that has to be employed for grafting or success would decline with the increase in maturity. At this stage scion stick should also be ready for which advanced preparations are necessary.
3. **Preparation of scion**: Select the tree of proven productivity and choice variety for grafting on stock plant. Choose more than three months old mature branches of appropriate thickness with dark green leaves and terminal bud. Defoliate the selected branch leaving only petioles 8 to 10 year prior to expected date of grafting. This period however may vary with the tree species. Due to defoliation the buds will start swelling and get activated. Such scion is ready for use. These scions can be cut in the afternoon and immediately used for grafting or wrapped in wet cloth and packed in polythene bag for taking for grafting. Such scion can be used even after 3 to 4 days.
4. **Method of grafting**: The top portion of the fresh growth on stock plant is decapitated keeping about 8 cm of tender shoot. Then a longitudinal cut of about 3 cm length is made from tip in this 8 cm fresh shoot on stock plant. A scion of almost same thickness of 10 cm length is selected for use. The lower end of the scion should be made to 3 cm as wedge by chopping off bark and little wood from two opposite sides, taking care to retain some bark on the remaining two sides. The wedge of scion so prepared should be inserted into the slit of the stock and secured firmly with 1.5 cm wide and 45 cm long 200 gauge white transparent polythene strip. Within 15 days union should be complete and plants will make growth. In plant species where appearance of new growth is slow, it can be forced through heavy heading back like in aonla and guava.
5. **After care:**
   a) One good plant should be retained at one spot after successful operation.
   b) New branches below the point of union should be regularly removed.
   c) In the event of death of grafted scion after union and sprouting, new sprouts from stock plants can again be utilized for grafting purpose at appropriate stage.
   d) In case the root stock has attained greater age (4 to 5 years), it can be headed back to appropriate level and new growth can be used for grafting at desired height.
   e) After the scion had sprouted and the union is perfect, the polythene strip should be removed or it will hinder the plant growth.

f) Fairly big size basin should be prepared around the plant and mulching be practiced during summer. If need be, in case of severe drought, spot irrigation can be given for one year for better and quick growth.

g) Necessary measures should be taken against disease and pest.

The technique is very efficient for the establishment of most subtropical and tropical in arid and semi arid condition.

B. Raising of Budding in Polythene Tubes

This technique was standardized for raising ber plants. Experience has revealed in ber that owing of seed in-situ is necessary because when raised in nursery seedling do not stand transplanting due to fast development of roots an their extensive nature which gets damaged due to lifting. Therefore, recommendation were made that seedlings can be raised in polythene bags of 9" x 4" size in April to May and such seedling could be transplanted in the field in July to August for in situ budding. Thereafter a technique was standardized for raising budding of ber polythene tubes of 25 cm length and 10 cm diameter. For this tubes are filled with a mixture of sand and FYM in equal amount by holding them in upright position in the nursery bed. Extracted seed of *Z. rotundifolia* and not the stones are sown in the tube during the first fortnight of April. These seed germinate within one week and in ninety days attain buddable size. It should be seen that seedlings are maintained on single stem through regular tending. The budded plants become ready for transplanting in August and the establishment of these plants its of very high order as roots are neither cut, as would be in case lifting from nursery, nor knotty as would be in the pots. It is very easy to transport budded plants in polythene tubes for short distance. However distant transportation has been causing some problem due to damage in transit for which a polypack method. In this technique the budded plant raised in polythene tube is taken out of the tube with soil ball. The soil around roots is washed out with water jet or dipping in thankful of water. The roots are then immediately wrapped in sphagnum moss saturated with nutrient solution of the following ingredient: Potassium nitrate 0.2 g/l, calcium nitrate 0.8 g/l, potassium hypophosphate 0.2 g/l, magnesium sulphate 0.2 g/l. Diathane Z-78 (0.2 per cent) is also added in the above solution. Excess of solution from the moss is squeezed out before it is used for wrapping the roots of the budding. The roots wrapped in moss are then put in a polythene bag keeping the shoots portion out. The open end of polythene is tied securely so as to avoid loss of moisture from the moss. The plants so packed keep fresh for 15 days and survive well after planting. One of these techniques can be used for raising fruit crops in arid and semi arid condition with maximum success.

Chapter 11

Morphological Features of Arid Fruit Crops

Distribution of fruit crops is determined almost entirely by climatic factor. The most important of which is temperature. In the hot arid zone of India temperature and humidity conditions are ideally suited for growing of fruits of high quality like Date palm, Ber, Guava, Fig, Olive, Pomegranate, Citrus, Papaya, Grapes *etc.* However there is hardly orchard in this region in spite of great potential. It is therefore necessary to select such fruit crops which can stand better under dryland condition.

Selection of Fruits

Strategy of development of orchard under dryland area the correct selection of fruit crop is of prime significance. The crop should have following attributes:

1. Fruit crop selection for dryland region must be such that with maximum growth period synchronized with the period of maximum availability of moisture and low vapour pressure deficit in the atmosphere. The ideally the period from flowering to fruiting must also fall during this period and fruit repining must be completed well before the onset of summer. Fruits like Ber, Custard apple, Phalsa, Lasoda, Fig, Wood apple can fit for this condition. Anthesis of crop should be synchronized with maximum moisture availability period in the atmosphere, *e.g.*, Guava, Pomegranate, Acid lime which bear fruit in distinct Bahar (flush) which considered with rainy season can only be encourage.
2. Since the water is the limiting factor in dryland. The crop should have drought tolerance and mechanism like-
   - **(a)** Crop should have deep tap root system and have the capacity to draw water from the deeper layer of soil profile, *e.g.* Ber.

(b) Shedding of leaf in summer to conserve the moisture like Ber and Lasoda this is very useful drought avoiding mechanism.

(c) Water binding mechanism as in Fig. In the tissue of Fig certain water binding chemical constituent are found which results high bound water.

(d) Other xerophytic leaf characters like wax coating, hairiness, reduce foliage, thick cuticle, sunken and covered stomata on the leaf. Minimum loss of water through transpiration as in Fig, Phalsa, Ber, and Lasoda.

3. Dessert soil are very poor in soil water holding capacity. They are also generally shallow and are with canker layer of 70-100 cm thickness at depth varying from 100-150 cm. Fruit crops capable of sticking their roots to penetrate this layer therefore the most adoptable to dessert layer crops is roots of *Zyziphus rotandifolia* (suitable rootstock which penetrate 100 cm thick canker layer.

4. Fruit crops of arid region should have tolerance to salinity and saline water which are common feature of these area and fruit crops like Aonla, Grapes, Ber have great tolerant to pH and can grow in the range of (9.2-10.5).

The Salt Tolerance of different Fruit Crops are as Follows

Fruits	*Salt Tolerance Limit*	
	ECE (mmhos/cm)	*pH/ESP*
Aonla	15	40 ESP
Ber	14	30 ESP
Grape	07	9.0 pH
Guava	08	9.0 pH
Karonda	—	10.0 pH
Mulberry	—	9.5 pH

5. Dry atmosphere and high heat accumulation is the hallmark of Indian arid zone. The crops suitable for such climate are: Date palm, Grapes and Papaya. Similarly intense radiation is another important feature of climate in arid region. The crop with good canopy to shade fruits and protect them from sun burning could be chosen like Kinnow mandarin which have tolerant against drought. As well as bear fruit inside well formed canopy.

6. Variety selected or involve for arid condition should be short duration *i.e.* early variety. So that they complete their fruiting period in shorted possible time to evade drought condition.

The fruit crops which have possibility of cultivation in dryland can be grouped into four classes:

(a) Indigenous drought hardy type: These are Ker (*Capparis apphylla*), Pilu (*Salvadora oleoides*), Lasoda (*Cordia myxa*) Gondi (*Cordia gharaf*), Ber (*Ziziphus spp.*)

(b) Fruit tree having reproductive phase synchronized with the period of moisture abundance *i.e. Carrisa carandus, Carrisa grandiflora*, Pomegranate, Custard apple, Guava *etc.*

(c) Fruit is a possibility of synchronized with the reproductive phase to the period of moisture abundance *i.e.* Sour lime, Aonla, Guava, Phalsa and Pomegranate

(d) De-irrigated fruit crops- fruit crops like Phalsa, Mulberry, Papaya, Fig, Grapes, Date palm, Mango, Sweet orange *etc.* these fruits would however invariably require irrigation facility.

Ber (*Ziziphus mauritiana*)

It is the most draught hardy fruit tree which can stand salinity and saline water. Its root can penetrate through hard pan and draw moisture from deeper layer of the soil. In this case flowering start with the onset of Monsoon and the crop is harvested well in advance of onset of hot and dry summer. Most interesting feature of this crop is that during hot summer it will shed off its leaves completely thus conserve moisture by gating rid of transpiration mechanism. At the same time the leaves have xerophytic character and the buds are scaly and thus stand intent summer heat with no damage.

Aonla (*Emblica officinalis*)

This crop has great tolerance to salinity, alkalinity and sodicity. About its ability to stand drought. It can be shed as if nature has created this plant only for drought condition. In this crop flowering take place in spring (March-April). After fruit set fruits inter dormancy and remain so throughout summer. Therefore plants do not require irrigation during summer. With the onset of Monsoon the fruit start growing and mature by December.

Guava (*Psidium guajava*)

This fruit crop is tolerance to soil salinity to certain extent and capable of growing to great variety of soil and climate. Guava tree can be tailored to only mrig bahar which coincide with the period of maximum availability of natural moisture. Thus the tree can be tailored to be flower in June- July from which crop can be harvested in December-January well before the onset of Hot summer. Leaving to nature them this tress shall shed off leaves and become absolutely leafless in summer.

Bael (*Aegle marmelos*)

It is drought and salinity tolerant plant in which flowering coincides with the onset of Monsoon and the fruit mature before the onset of hot summer. It can stand swampy and alkaline soil as well. A grown up tree yield about 300-400 fruit per year, which hardly have any table value but it make excellent squash or beverage. The Bael is considered to have many medicinal properties and is effective in the treatment of dysentery.

Wood Apple (*Feronia lemonia*)

It is a tree of rutaceae family and has great tolerance to drought and salinity. Extensive root system and synchronization of its reproductive phase with high moisture availability make it a suitable crop for arid region. Fruits hardly have any table value but the pulp is extensively used for chutney preparation and being rich in pectin. It can be an excellent source in jelly preparation and the tree is excellent for waste land.

Jamun (*Syzygium cumunii*)

Ones established the plant this can stand drought which can be attributed to extensive root system. Flowering start in spring and the final swell of fruit takes place in Monsoon. Its fruits ripen a little later in a season than Mango.

Karonda (*Carissa carandas, Carissa grandiflora*)

The karonda plant is drought tolerance and its reproductive phase synchronized to the period of moisture abandons. Therefore it is highly suitable for arid region. Fruits are small and acidic and highly price for their pickling value, excellent jelly and candy.

Date Palm (*Phoenix dactylifera*)

Date palm is dioecious in nature and has great tolerance to salinity and saline water irrigation. This crop require sufficient heat unit by the end of June before onset of Monsoon and high vapour pressure deficit during fruit development period. These conditions are require for faster fruit development and avoidance of damage to fruit due to cracking and splitting which may take place due to high humidity and rains.

Fig (*Ficus carica*)

Plants can great tolerance to salinity and drought. Fruiting can be monitored to synchronize with rainy season only so that crop is harvested in later winter and plants will shed off leaves during summer and overcome the drought condition.

Pomegranate (*Punica granatum*)

Fruit crop can great tolerant to salinity as well as saline irrigation water and grow even in shallow soil. It can also tolerant to drought as fruiting can be tailored to synchronized with maximum availability of moisture in the atmosphere. Under normal condition Pomegranate flowers three times in a year. *i.e.* Ambe bahar (spring flowering), Mrig bahar (rainy season) and Hasth bahar (post monsoon season). If we concentrate only on Mrig bahar and keep tree under stress during the rest of the period it will adjust its growth and fruiting period to arid situation and plant will even shed off leaves during summer to evade water loss.

Custard Apple (*Annona squamosa*)

This is also a very hardy fruit crop have great tolerance to drought salinity and saline irrigation water to some extent. It also grows very well on shallow soil fruiting coincide with maximum moisture availability period with flowering in June-July and harvesting by November-December. It also shed off leaves during hot and dry summer to evade moisture loss from plant tissue through transpiration.

Phalsa (*Grewia subinaequalis*)

Its belong to tiliaceae family and plant is hardy and has a very short reproductive phase and after the harvest of the crop in early summer the plant tend to lose foliage to evade stress of sever hot summer. Flowering start in December-January and fruits ripen in April.

Gonda/Lasoda (*Cardia myxa*)

The most interesting feature of gonda tree is that after the crop is over in early summer the tree shed off leave and stand severe summer without any loss or mortality. The flowering take place in December-January and crop is ready within three month. Since the fruits are used only for pickling. They are harvested in immature stage.

Acid Lime (*Citrus aurantifolia*)

It is relatively drought tolerance among citrus species and can grow in shallow soil. It can be made to tolerate soil salinity and saline irrigation water by employing salt tolerant rootstock. Once established plant can be talered to flower during Monsoon and major crop would be ready for harvest in late winter well before the onset of summer.

There are many indigenous drought hardy plants which even exploited can be of commercial value these plants have drought hardy characteristics like deep root system, (Tamarind: *Tamarindus india*), scanty foliage (Kair: *Capparis decidua*), Musilagenous sap in plant part which will hold water and not loose it too during summer months (Kair: *Capparis decidua*), (Gonda: *Cordia myxa*), (Pilu: *Salvadora oleoides*) and (Bael: *Aegla marmelos*). These trees are found fruiting profusely even during extremely hot summer and under scanty rainfall.

Beside these there are many high value fruit crops like grapes, sweet orange, Papaya, Mango, Kinnow and Mulberry. Which require dry weather during flowering and fruiting for developing high quality fruits. Which the provision of irrigating facility these crops can be raised for high quality fruits.

Section-II

Cultivation Practices of Arid Fruit Crops

Chapter 12

Aonla

Botanical name:	*Emblica officinalis*
Family:	Euphorbiaceae
Chromosome no.:	2n = 28
Origin:	Tropical South-Eastern Asia
	Particularly Central and Southern India

Introduction

Aonla or Indian gooseberry is also known as Indian amla. Aonla is in cultivation in India since time immemorial. Its recital are found in Vedas, Ramayan, Charak Samhita. Sushrut Samhita, literature of Kalidas, Kadambari and other ancient literature. It is more popular as a backyard fruit throughout the country. Aonla is the native of tropical south-eastern Asia particularly central and southern India. It is commercially grown in U.P. and Gujarat. In Uttar Pradesh aonla cultivation is more popular in Pratapgarh, Mirzapur and Barely.

Taxonomy and Morphology

Aonla is a tree of medium height. It is grouped as evergreen in tropics but in subtropics it exhibits deciduous nature due to abscission of determinate shoots. Phyllanthoid branching habit with two types of shoots; determinate (short shoots) and indeterminate (long shoots). The leaves are so dischously arranged on determinate shoots that they look like pinnately compound leaf. However, the leaves are simple 1 to 2 cm long. Flowers are borne in the axil of the leaves on determinate shoots as axillary cymules. Basal cymules are of male flowers followed by nodes each with a cymule of one central female flower (rarely two) and several lateral males. A typical flower consists of parianth and androecium or gynoecium. Number of parianth segments varies from 5 to 7 but generally 6. Female flowers take 72 hours to open. Pedicel is very short. Disc is lanceolate cup with 3 carples. Style is short, connate, twice bifid, and distally dialated. Flowering takes place in spring. After fruit set the fruits remain dormant through summer and start swelling with the onset of monsoon and ready for harvest in December/January depending on the location.

Fruit is capsular (drupaceous) with a fleshy exocarp. The endocarp (cocci) is rather thick and massive. Eventually, separating from a well developed Collumella. Collumella is trigonous and slender below the dialated apex. Seeds are two in each locule and a seed coat is dark chestnut brown and hard.

Composition and Uses

Aonla fruit is very rich in vitamin C and pectin, therefore, regarded very important for medicinal value in Ayurvedics. Undoubtedly, it is a richest natural source of vitamin C known to mankind. A tannin containing gallic acid, ellagic acid and glucose in its molecule, which is naturally present in the fruit, prevents or retards the oxidation of vitamin C and renders it a valuable anti scorbutic in the fresh as well as in the dried and processed condition. This fruit is useful in haemorrhages, diarrhoea, dysentery, anaemia, jaundice, dyspepsia and cough. It. is an important ingredient of Triphala and Chavanprash in Ayurvedic medicine system. Fruits are commonly used for preserve (murabba), pickle, candy, jelly, jam *etc.* It can be dried and powdered to be used subsequently. Used in the preparation of inks, hair dyes, hair oils. It is a great health and vitality restorer.

Climate and Soil

It can be grown successfully in arid climate also called as king of arid fruit. This is a fruit of subtropical region with distinct winter and summer. However, in India it is being grown near sea coast up to 1800 m altitude. Soon after fruit set in spring, the fruits remain dormant through summer without any growth, makes it highly suitable fruit crop for arid region.

It is not very exacting in its soil requirement and grows well in sandy loam to clay soils in India. It has great tolerance to salinity and sodicity and cultivated in pH range of 6.0 to 8.0 very successfully. However, production shall be highly benefited in deep and fertile soils.

Important Varieties

Most of the plantations have been raised from seed, therefore, exhibit great variability and the fruits are of inferior quality. However, some named cultivars like 'Banarasi', 'Francis' (Hathijhool), 'Chakaiya' and 'Bansi Red' are known to be cultivated in U.P. 'Krishna and 'Kanchan' are new finds from Ayodhya U.P. and 'Anand-1' and 'Anand-2' from Gujarat from seedling variables available in the area.

- **Banarasi:** Erect, spreading, medium to tall tree. Shy and slightly alternate bearer. Fruit is roundish weighing about. 50 g. whitish green, slightly fibrous, scanty to medium juicy.
- **Francis (Hathijhool):** Erect and tall tree, good and regular bearer. Large fruited weighing about 63 g, oval roundish, light green, slightly fibrous, moderately juicy.
- **Chakaiya:** Spreading, medium height, prolific and regular bearing tree. Medium size fruit, flattened at base and round apex, greenish colour and fibrous.
- **Kanchan (NA-4):** Chance seedling of chakaiya fruit is small to medium, flattened oblong, skin smooth, light green segment six and difficult to separate. Flesh fibrous and hard, keeping quality is good and ideally suitable for preparation of pickles.
- **Krishna (NA-5):** This is the chance seedling of Banarasi from Pratapgarh district of Uttar Pradesh. Fruits are large triangular skin smooth, whitish green to apricot yellow in colour, segment six easily separated less fibrous and highly astringent.
- **NA-6:** It is a selection from open pollinated chance seedling of chakaiya, fruits are small to medium and oval in shape, skin smooth, semi translucent and light green in colour, flesh is fibrous and semi hard, seed are small light brown in colour, keeping quality of fruit is good and suitable for processing. Fruits are free from internal necrosis.
- **Neelum (NA-7):** It is a seedling selection from open pollinated strain of francis. Fruit are medium to large conical in shape, skin smooth, semi translucent, yellowish green, segment six and easily separated, flesh almost fibreless and soft. It is free from fruit necrosis. Hence can be used for processing.
- **NA-8:** It is selection from open pollinated seedling of chakaiya, fruits are small and flattened round, skin slightly rough, thick and light green in colour. Flesh is fibrous in colour. Keeping quality is good, it is susceptible to fruit necrosis.
- **NA-9:** It is seedling selection from open pollinated strains of cultivar Banarasi, fruit large in size, flattened round, skin smooth, light green in colour, flesh soft and moderately fibrous. Shy bearer, early maturing and having moderate keeping quality. Fruit is susceptible to fruit necrosis.
- **NA-10:** It is chance seedling develop from cultivar Banarasi and locally known as Agra Bold. Fruits are flattened round and medium in size, skin is rough, having yellowish green colour, flesh slightly fibrous, whitish green, soft juicy and highly astringent. Keeping quality is good, fruits are moderately susceptible to fruit necrosis.
- **Goma Aishwarya:** Goma Aishwarya is a selection from the tree identified from the large population. It is an early drought tolerant variety and

prolific bearer, tree is semi –spreading and fruits are medium in size. It has low fiber content and is suitable for processing and export.

Propagation

Aonla is propagated by seed as well as by vegetative methods. Since seed multiplication brings variability, vegetative methods should only be adopted. However, for raising root stock, seedlings are required to be raised through seeds. Seeds are collected from fully matured fruits and sinkers are used for sowing and floaters discarded being embryo less. There is no seed dormancy, therefore, fresh seeds give almost 100 per cent germination. Process of germination can be quickened by soaking dried seeds in 500 ppm GA solution for 24 hours. Seedlings can be raised in seed beds or polythene bags. It will take them about 4 months to attain buddable size. Both inarching and budding methods can be used for raising plants but budding is more successful. Among three methods of budding attempted, Forkert and Patch budding done from June to September, gave maximum success.

However, T-budding gave 80 per cent success in rejuvenation of old orchards Even soft-wood grafting can also be used successfully for which seedlings are raised in situ and headed back severely in the month of May for forcing new vigorous side shoots. On these new shoots, soft wood grafting method can be employed by way of wedge grafting. Use of new apex of root stock and activated scion is necessary.

Planting

Grafts or budding of aonla are best planted in the beginning of monsoon in the months of June to July. Since the tree grows to a huge size a distance of 8 to 10 m both ways is recommended. In areas with irrigation facilities, planting can also be done in spring (February to March). Before planting, it is necessary that planting place is appropriately marked, 1x1x1 m size pits are made well in summer, kept open for about a fortnight. In each pit 3 to 4 baskets of FYM are mixed with dug soil and filled. After first rain the plants are planted in the centre of these pits and staked properly.

Training and Pruning

Aonla tree does not require regular pruning. However, pruning in early years for giving proper shape and development of strong frame may be necessary for which tree should be trained to single stem up to the height of about 1 m and then primary branches can be allowed at regular space all around the trunk.

Orchard Soil Management

It refers to management of soil in such a way that productivity is maintained and soil losses are minimum. The technique should be compatible to crop behaviour. In aonla flowering takes place in February and after fruit set the fruits remain dormant through summer until monsoon when fruits begin to grow and ready for harvest in December/January depending on the climate of the place. Therefore, in summer, soil should not be disturbed and natural cover can be allowed. In the beginning of monsoon the basin can be cleared, fertilizer and manurial application can be affected and the place between rows could be brought under green manure to be ploughed in at flowering stage and left as such.

Nutritional Management

Regular orcharding of aonla is a rare phenomeson, therefore, nutrition is hardly However, beneficial effect of nitrogen, phosphorus, potash, zinc, copper, manganese and boron have been found beneficial. Nitrogen increased vegetative growth and female flowers, while P increased sex ratio, initial set, fruit retention and yield, T.S.S. and vitamin C content of the fruit. Potassic fertilizer increased the fruit retention and quality. A bearing ten year old tree should be given 1.5 kg N, 1.0 kg K and 0.7 to 0.8 kg P annually under the drip of the tree where feeding roots are mostly present. 0.6 per cent borax thrice in the months of September and October at about 10 to 15 days interval to control fruit-necrosis which develops due to the boron deficiency.

Irrigation

Aonla trees are hardy and stand very well against drought. Therefore, hardly any irrigation is practiced. However, the crop shall be benefitted by giving two/ three irrigations at the time of full bloom and set. During summer, when the fruit is dormant, there may not be any benefit to irrigate trees.

Flowering and Fruiting

Flowering takes place on determinate shoots appearing in spring season and the initial percentage of fruit set may vary from 12 to 18. Pollination is helped by honeybees. Initial poor set is due to low sex ratio and lack of pollinators or pollinizer or both. Soon after fertilization, which takes place within 36 hours of pollination, the zygote rests for 120 to 130 days and endosperm nucleus may rest for 70 to 80 days after fertilization in Banarasi. Therefore, fruits do not show any symptoms of growth during summer. Dormany breaks by the end of July as a result of division in endosperm nucleus. Fruits begin to grow with maximum growth in September and attain maturity by the end of November. During this growth phase, flowers and fruits also undergo abscission which has been divided into three ways:

a. First drop consists of unfertilized ovaries and degenerated ovules.

b. The second drop consists of drop of young fruit lets at the time of dormancy break.

c. Third drop is spread over a period of rapid growth from August to October may be due to lack of auxins.

Insect Pest and Diseases

Insect Pest

1. **Gall caterpillar (*Betousa stytophora*):** Young caterpillars bore into the apical portion of the shoot during rainy season and make tunnel. Due to this, damaged region bulges out abruptly into a gall which provides space for movement of the caterpillar. Due to this apical growth is checked, side shoots develop below the gall and subsequent growth in following season is greatly hampered. This can be controlled by cutting away the infected apices and prophylactic spray of systemic insecticide like Rogor 0.03 per cent.
2. **Leaf rolling caterpillar (*Garcillaria acidula*):** This caterpillar rolls the leaf and feed inside reducing the photosynthetic capacity of leaves and causes subsequent leaf shedding. It can be controlled by spraying 0.08 per cent Malathion or 0.04 per cent Monocrotophos.
3. **Bark eating caterpillar (*Inderbela tetraonis*):** It damages stem and branches of grown up trees by eating bark. This insect is the problem of neglected orchards. Affected portion should be cleared of brown dust and a few drops of kerosene oil or petrol should be applied in holes to keep this in control.
4. **Mealy bug (*Nipaecoccus vastator*):** Both nymphs and adults are reported to feed on aonla tree from April to November. Organophosphates provide excellent control of this insect. Monocrotophos 0.04 per cent or Malathion 0.08 per cent or methyl parathion 0.03 per cent are effective as sprays.

Diseases

1. **Ring rust (*Ravenelia emblicae*):** Ring rust appears as circular or semi-circular, reddish solitary or gregarious spots on leaves from the beginning of August. Infection on fruits follows. Generally, one or two pustules measuring 10 to 20 mm in diameter appear on infected fruit. Occurrence can be prevented by spraying 0.2 per cent Dithane Z-78 at the interval of 7 to 28 days during the months of July to September.
2. **Fruit rot (*Penicillium oxalicum P. islandicum; P. oxalicum* and *Aspergillus niger*):** A number of organisms are associated with this rot in which a few affected fruits are found in the orchard. However, the major loss takes place during transit to the market. The earliest symptom of infection is seen as water soaked lesion on the fruit surface, which enlarges in size followed by development of small pin head size colonies of golden yellow colour. The older colonies turn olive green. It is recommended that fruits showing such symptoms are discarded for marketing. Bruising and injury at the time of harvesting should be avoided as a preventive measure. Treatment of fruits with borax and NaCl to control the diseases.

3. **Leaf rust (*Phakospora phyllanthi*):** This rust is commonly seen on leaves in the months of July and August and Bordeaux mixture 0.1 per cent will control this.

Physiological Disorder

Internal fruit necrosis: In this disorder necrosis symptoms start with the browning of innermost part of the mesocarpic tissues at the time of endocarp hardening in the second and third week of September. Browning of mesocarp extends towards the epicarp resulting into brownish black areas on the fruit surface in the second and third week of October. Depending upon the severity of the disorder, mesocarp of affected fruit turns black and becomes corky and gummy pockets develop. This has been reported to be due to the deficiency of Boron for which spray of 0.6 per cent Borax thrice in the month of September and October at about 10 to 15 days interval are recommended to control the disorder.

Harvesting and Yield

A vegetatively propagated tree starts fruiting commercial crop after 6 to 8 years of planting, while seedling trees may take 10 to 12 years to begin bearing. Productive life of trees is estimated to be 50 to 60 years under good management.

Generally aonla fruits are ready for harvest in November/December. Their maturity can be judged either by the change of seed colour from creamy white to black or by the development of translucence exocarp. Maximum vitamin C content is observed in mature fruits while immature fruits are acrid and low in vitamin C content and minerals. The production varies from cultivar to cultivar. 'Banarasi' is a poor yielder as compared to 'Chakaiya', and 'Francis'. On an average a grown up tree should yield 100 to 150 kg per annum.

Chapter 13

Bael

Botanical name:	*Aegle marmelos*
Family:	Rutaceae
Chromosome no.:	2n = 18
Origin:	India

Introduction

Bael is one of the oldest fruit cultivated in India from prehistoric times. It is also grown in Sri Lanka, Pakistan, Bangladesh, Burma and Thailand. In India, it has great mythological significance and therefore, found growing near Shivalayas. Most of the spiritual writings have mention of this fruit. This may probably be due to great therapeutic value of various parts of this tree. However, no systematic orcharding is known except a few in U.P. But in recent years, due to its ability to tolerate arid condition and processing quality, it is receiving umpteen attention from scientists for improvement.

Taxonomy and Morphology

This is a fruit of genus Aegle which belongs to the great citrus family Rutaceae and consists of 2 or 3 species. The tree is deciduous grows to great heights with white-greyish stem. Leaves are trifoliate, branches are spiny, and bark is shallowly furrowed and corky. Flowers are bisexual borne in clusters sweet scented and greenish white. Stamens are numerous sometimes coherant in bundles. The ovary is oblong-ovoid, slightly tapering with numerous ovules in each cell of 8 to 20 in each ovary. Fruit is a hard shelled berry usually globose with a pericarp nearly smooth greyish yellow about 3 mm thick, hard and filled with soft yellow and orange, very fragrant and pleasantly flavoured pulp. Seeds are many com- pressed and arranged in closely packed tiers surrounded by slimy mucilage.

Composition and Uses

The bael fruit is highly nutritious. 100 g edible pulp will contain 61.5 g water, 1.8 g protein, 0.39 g fat, 1.7 g minerals, 31.8 g carbohydrates, 55 mg carotene, 0.13 mg thiamine, 1.19 mg riboflavin, 1.1 mg niacin and 8 mg vitamin C. A report from Brazil indicates 65 to 100 mg vitamin C per 100 g pulp. It has numerous uses.

1. From the pulp sherbet and syrup can be made. The marmalade prepared from its fruits is used in diarrhoea and dysentery. It can also be made into preserve.
2. Fruits give mucilaginous substance, used as a cement mixture with lime.
3. Gum is obtained from stem.
4. The wood is used for making agricultural implements.
5. The twigs are used as firewood.
6. The leaves are used to worship 'Lord Shiva' according to Hindu custom.
7. The leaves are also used as fodder.
8. All the parts of the plant are medicinally important due to marmelosin.

Climate and Soil

The bael trees grow successfully and produce higher yields in sub-tropical climate where summer is hot and dry and winter is mild. Further, bael tree is reported to be hardy and can be grown up to an altitude of 1219 m and is not damaged by temperature as low as -7° C.

The bael tree is very hardy, it can thrive well even in swampy, alkaline and stony soils having pH range from 5 to 10, where many other fruit trees fail. However, well drained sandy loam soil is the best.

Important Varieties

Among local type Ayodhya Kagzi, Etawah, Gonda and Mirzapuri bael are popular in Uttar Pradesh in recent year some promising variety like Pant Bael-3, Pant Bael-1, Pant Bael-9 (Pant Aparna) and Pant Bael-12 (Pant Sujata) produce dwarf tree which are develop from Govind Ballabh Pant University of Agriculture and Technology, Pant Nagar. Pant Bael-10 and Pant Bael-1 produce higher number of fruit per tree while Pant Bael-5 (Pant Shivani) fallowed by Pant bael-12 Produce highest yield in term of fruit weight per tree.

NB-5, NB-7 and NB-9 are another promising cultivar which performs well under arid condition is developing by Narendra Dev University of Agriculture, Kumarganj Ayodhya. Recently CISH-B-1 and CISH-B-2 selection of Bael have developed at Central Institute of Sub-tropical Horticulture, Lucknow.

Propagation

Bael can be propagated by both the method sexual and asexual in sexual method. For raising seedlings, the seeds are sown in the nursery bed. The seeds have no dormancy and they germinate in 3 to 4 weeks after sowing. The germination of seed hypogeous, the young seedlings should be protected from frost.

Some root suckers which arise from the roots of the bael trees can be separated during monsoon and planted. Patch budding proved to be the best when carried out in the month of June or July. Propagation of bael by grafting is also successful. It can be grafted on to *Aegle fraeglegabonensis* and *Aeglopsis chevalieri*.

Planting

The pits of 50x50x50 cm are dug and filled with top soil mixed with 10 to 15 kg farm yard manure. The distance of planting is 10 x 10 m to 12 x 12 m. planting can be done in February to March and June to July. However, for arid conditions beginning of the monsoons is the most appropriate period.

Training and Pruning

It hardly needs any training or pruning. Budded plants need staking in the beginning to grow erect. Lateral branches up-to the height of one metre should not be allowed to grow. Zig zag or criss cross branches should be removed to facilitate easy penetration of air and sunlight inside the canopy.

Nutritional Management

No systematic work has yet been taken up on nutritional requirement of bael tree. Occasionally, cultivators do shallow cultivation and apply some organic manure. Since plant is hardy and produces ample number of fruits, therefore, people might have not bothered about fertilization. However, it will be benefited from nutrient application and an application of 20 to 30 kg FYM/annum is recommended in the beginning of monsoons. If it is supplemented with 500 g nitrogen per plant in two split doses of 250 g each in February and June, it will help in increasing yield and improve fruit quality.

Irrigation

Young plants need to be cared for watering time to time in the first year of planting and when plants establish well, usually watering is not done. However, sufficient moisture helps in good yields for which runoff collection should be practiced in rainy season by developing micro catchments around the trees.

Flowering and Fruiting

The seedling trees require 7 to 8 years to bear fruits while budded plants start bearing fruits at the age of 4 to 5 years after planting. Bael tree flowers in May to June and fruits become ready after 8 to 10 months or so say by April to May. Bael fruit has three distinct periods of its growth. The initial growth is slow for about one month followed by a rapid increase for four months and then more or less a stationary phase until the fruits are harvested. Based on respiratory studies bael fruit can be classified as a climacteric fruit

Insect Pest and Diseases

Insect Pest

No serious pest or diseases have been noticed on bael

Disease

Shot-hole and fruit canker (*Xanthomonas bilvae*): This is a bacterial disease. Affected leaves show round, water soaked spots surrounded by a clear halo. Later on spots increase and form brown lesions. These lesions fall down resulting in shot holes. The causal organism also infects the fruit, twigs, and thorns.

Control

a. Remove affected portion from the tree.

b. Spray bactericide (antibiotics)

Physiological Disorder

Fruit Cracking- The fruit cracking is a major physiological disorder and its degree of damage depends according to genotypes/varieties and locality. The fruit cracking takes place twice in a year *i.e* i) during winter season (December-January) while developing fruits are immature, ii) during summer season (March-April) when fruit are mature and in ripening phase. The cracking at later stage is more severe than former one. The cracking can be minimized by maintaining optimum soil moisture regime and by creating of wind breaks against hot desiccating wind side of orchard. Organic mulches like paddy straw, maize straw, *subabul* loppings can also be effectively used in maintaining soil moisture of tree basin particularly during summer under rainfed conditions of hot semi-arid ecosystem. It is due to nutritional deficiency, especially boran and calcium. Control of cracking of fruit spray of borax @ 0.1 per cent twice at full bloom and after fruit set.

Fruit Drop- Fruit drop is a natural phenomenon, but its extent of damage is a matter of concern. The extent of fruit drop varied according to genotypes/ varieties and locality. Immature fruit drop (marble size) has also been observed. Sometimes cricket ball size fruits also fall down during August. The extent of fruit drop in bael can be reduced effectively by adopting better orchard practices which include mulching with organic materials and proper soil nutrient management (macro–micro) and application of growth hormones like NAA (15-20 ppm/litre) at pea size stage during August-September.

Harvesting and Yield

Mature bael fruits are harvested individually from the tree along with a portion of fruit stalk. To prevent fruit from falling on the ground fruit picker is used for harvesting. Bael fruits are climacteric and therefore, should be harvested at maturity. Fully ripe fruits on trees start cracking if harvesting is delayed. At harvesting precautions should be taken to avoid falling of fruits on ground. In north India, fruit ripens in April and May while it ripens a bit earlier, in eastern India.

A 10 years old tree yields about 100-150 small sized fruits or 50-75 big sized fruits. A well grown up tree of about 12 to 15 years age produces 300 to 500 fruits under good management.

Chapter 14

Barbados Cherry

Botanical name: *Malphigia punicifolia* L.

Family: Malpighiaceae

Chromosome no.: 2n = 40

Origin: Trinidad and Tobago

Introduction

West Indian cherry, also known as Barbados cherry is the richest source of vitamin C. It is a medium sized shrub, which thrives well in tropical climate. It is best suited as a homestead fruit crop and prefers a rich well drained soil. The Barbados cherry, a member of the Malpighiaceae, is an interesting example of a fruit that rose, like Cinderella, from relative obscurity about 40 years ago. The Barbados cherry is a large, bushy shrub or small tree attaining up to 20 ft (6 m) in height and an equal breadth; with more or less erect or spreading and drooping, minutely hairy branches, and a short trunk to 4 in (10 cm) in diameter. It is the evergreen fruit crop.

Taxonomy and Morphology

It was at that time the subject of much taxonomic confusion, having been described and discussed previously under the binomial *Malphigia punicifolia* L., which properly belongs to a wild relative inhabiting the West Indies, tropical America and the lowlands of Mexico to southern Texas, and having smaller, pointed leaves, smaller flowers in peduncled umbels, styles nearly equal, and smaller fruits. *M. punicifolia* L. (*M. glabra* Millsp.) has been generally approved as the correct botanical name for the Barbados cherry, which is also called West Indian cherry, native cherry, garden cherry, French cherry.

Composition and Uses

Barbados cherries contain calories 59 g, moisture (81.9-91.10 g), carbohydrates (6.98-14.0 g), and ascorbic acid (vitamin C) (1500-4515 mg) per 100 gram fruit. Wine made from Barbados cherries in Hawaii was found to retain 60 per cent of the ascorbic acid. Barbados cherries are eaten out-of-hand, mainly by children. For dessert use, they are delicious merely stewed with whatever amount of sugar is desired to modify the acidity of the particular type available. Other-wise, the cooked fruits must be strained to remove the seeds and the resulting sauce or puree can be utilized as a topping on cake, pudding, ice cream or sliced bananas, or used in other culinary products. Commercially prepared puree may be dried or frozen for future use. It can be used for gelatin desserts, punch or sherbet, and has been added as an ascorbic acid supplement to other fruit juices. The fruits may be made into syrup or, with added pectin, excellent jelly, jam, and other preserves. The pasteurization process in the canning of the juice changes the color to orange-red or yellow, and packing in tin cans brings on further color deterioration.

Climate and Soil

The barbados cherry can be classed as tropical and subtropical. The mature trees can survive brief exposure to -2.22 ^{0}C. Young plants are killed by any drop below -1.11 ^{0}C. It is naturally adapted to both medium and low rainfall regions; can tolerate long periods of drought, though it may not fruit until the coming of rain.

The tree does well on limestone, marl and clay, as long as they are well drained. The pH should be at least 5.5. Elevation to 6.5 significantly improves root development. Acid soils require the addition of lime to avoid calcium deficiency and increase yield. The lime should be worked into the soil to a depth of 8 in (20 cm) or more.

Important Varieties

Two distinct types are seen.

Pink flower type: Flowers are pink and are born in clusters in leaf axils. Fruits are large in size (about 6 g) and red coloured when ripe.

White flower type: Flowers are white and are borne in clusters in leaf axils. Fruits are small (about 1 g) and orange coloured when fully ripe.

Propagation

If seeds are used for planting, they should be selected from desirable clones not exposed to cross-pollination by inferior types. They should be cleaned, dried, and dusted with a fungicide. It should also be realized that the seeds in an individual fruit develop unevenly and only those that are fully developed when the fruit is ripe will germinate satisfactorily. Germination rates may be only 50 per cent or as low as 5 per cent. Seedlings should be transferred from flats to containers when attained 5-7.5 cm height.

Air-layering (in summer) and side-veneer, cleft, or modified crown grafting are feasible but not popular because it is so much easier to raise the tree from cuttings. Cuttings of branches 1/4 to 1/2 in (6-12.5 mm) thick and 8 to 10 in (20-25 cm) long, with 2 or 3 leaves attached, hormone-treated and set in sand or other suitable media under constant or intermittent mist, will root in 60 days. They are then transplanted to nursery rows or containers and held in shade for 6 months or a year before being set out in the field. Some fruits will be borne a year after planting but a good crop cannot be expected until the 3rd or 4th year. The tree will continue bearing well for about 15 years. There is a lapse of only 22 days between flowering and complete fruit maturity.

Planting

For planting, pits of size 0.5 m x 0.5 m x 0.5 m are made, 6 m apart. Fill the pit with topsoil and 10 kg cow dung. After planting, mulching may be done with dry leaves to conserve moisture. July to December is considered to be the best time for planting. Irrigation once in four days during early stages of growth up to one year of planting and later on, once in 7-10 days is necessary.

Training and Pruning

Training is not required for the fruit plants. Pruning is done once in a year to maintain regular shape. Pruning consists of removal of dried and diseased wood and also the drooping branches.

Nutritional Management

A fully grown, bearing plant should be top dressed with fertilizers @ 100 g N, 160 g P_2O_5 and 260 g K_2O. These may be applied in two splits, in June-July and again in January, when there is sufficient moisture in the soil.

Irrigation

Young plants need to be cared for watering time to time in the first year of planting and when plants establish well, usually watering is not done. However, sufficient moisture helps in good yields for which runoff collection should be practiced in rainy season by developing micro catchments around the trees.

Flowering and Fruiting

Rooted cuttings and layers flower in six months. Flowering commences in the middle of May and extends up to August. The harvest of fruits commences from August and continues up to November. Rarely, flowering is noticed in March and the crop comes to harvest in April/May.

Insect Pest and Diseases

Insect Pest

One of the major obstacles to successful cultivation of the Barbados cherry is the tree's susceptibility to the root-knot nematode, *Meloidogyne incognita* var. *acrita,* especially in sandy acid soils. Soil fumigation, mulching and regular irrigation will help to keep this problem under control. The burrowing nematode, *Radopholus similis,* is also a cause of decline in otherwise healthy trees.

Diseases

Anthracnose (*Colletotrichum gloeosporioides)* and Leaf spotting (*Cercospora bunchosiae)* are disease found in the cherry fruit which are controlled by the Bavistin 2 g/lt water.

Harvesting and Yield

For home use, as dessert, the fruits are picked when fully ripe. For processing or preserving, they can be harvested when slightly immature, when they are turning from yellow to red. As there is continuous fruiting over long periods, picking is done every day, every other day, or every 3 days to avoid loss by falling. The fruits are usually picked manually in the cool of the early morning, and must be handled with care. Harvested fruits should be kept in the shade until transferred from the field, which ought to be done within 3 hours, and collecting lugs are best covered with heavy canvas to retard loss of ascorbic acid. There is great variation in productivity. Individual trees may yield 13.5-28 kg in Puerto Rico. In Jamaica, maximum yield in the 6th year is about 36 kg per tree; 24,000 kg/ha. 'Florida Sweet' in Florida has yielded 65 tons/ha.

Chapter 15

Ber

Botanical name:	*Ziziphus mauritiana*
Family:	Rhamnaceae
Chromosome no.:	2n = 24
Origin:	Indo-china region

Introduction

Ber is one of the ancient fruit of India and china. Its fruits are eaten fresh as well as dry and processed into delicious candy. It is one of the most important fruits of arid and semi-arid region. Although mainly a crop of arid and semi-arid region it is even growing exceedingly humid region of costal area including the Andmon Island. Some of the xerophytic characters and its ability to tolerate drought make it as a king of fruit of arid and semi-arid. The important characters are as fallows.

1. Deep and extensive tap root system capable to penetrate through hard pan and draw water from deeper layer of soil.
2. Ability to become quiescent during summer which is characterized by shedding off leaves and growth cessation.
3. Dark tan color and hard and pubescent scale develop on the bud resulting in to reduction of dehydration of bud.
4. Ability to complete fruiting phase during the period of moisture availability.

Due to these adaptive mechanisms it's highly shoots to arid and semi-arid condition and can be a very useful plant for economic up-liftment of people in such areas.

Composition and Uses

Ber fruits are very nutritious and rich in vitamin C, vitamin A and B complex and sugar. Ber fruits are also rich in menials like Ca, Mg, Iron, and Zink. ber fruits are eaten fresh as well as dried and processed in to delicious candy, ber jam, pulp *etc.*

Climate and Soil

Ber is found growing under tropical and sub-tropical climate all over India and thrive well up to an altitude of 1000 m above sea level. It is sensitive to freezing temperature but can withstand steamily hard condition by shedding leaf and interring it to quiescence. In north India growth start from late June to early July with the advent of Monsoon and fruit mature during February to April. In western India Monsoon set in early and the winter is mild and warm. Therefore fruit maturity in earlier during December to March. In south India flowering take place in May to July and fruit mature in October to November. The ber prefer atmospheric dryness for high quality fruits.

Soil is no limiting factor for ber cultivation. It can be grown in wild variety of soil ranging from shallow to deep and from gravelly and sandy to clay soil. It can tolerate salinity, alkalinity and even water logged condition to certain extent.

Species

The genus *Zizyphus* consists of 40 species in tropical and subtropical regions. *Zizyphus jujuba* (Cheneese ber) and *Z. mauritiana* (Indian ber) are the two species grown widely in temperate, tropical and subtropical regions of the world respectively. Beside *mauritiana* other important species like *Zizyphus rotundifolia, Zizyphus numuleria, Zizyphus rugosa, Zizyphus xylopyrus, Zizyphus micropyla, Zizyphus sativa* is also commercially grown for rootstock purpose.

Important Varieties

Early fruit ripening cultivar namely Seb, Gola and Mundia, perform better in arid area where as rain fed crop is taken fruit ripening is over by the end of February well before the rise of vapour pressure deficit in atmosphere. The cultivar Umran though successful throughout the country does not perform well under stream arid condition due to late ripening. However where the average rainfall is more than 500 mm. irrigation facility are available Umran can be grown successfully. Recommended cultivar for the important ber growing zone of the country are as follows.

Sl.No.	Name of the State	Suitable Varieties
1	Gujarat	Umran, Gola Mehrun and Randary
2	Haryana	Gola, Kaithli, Banarasi Karaka, Umaran, Seo
3	Maharashtra	Mehrun, Darakhi, Kharki, Shamber
4	Punjab	Gola, Najuk, Seo, Sandhura, Narnaul, Kaithli Banarasi, Umran
5	Rajasthan	Gola, Seo, Jogia, Mundia, Katha Phal

Sl.No.	Name of the State	Suitable Varieties
6	Uttar Pradesh	Narma Varanasi, Delhi Gola, Banarasi Gola, Banarasi Karaka, Mundia, Aliganj, Jogia, Muriya
7	Bihar	Nagpur, Banarasi, Harnless
8	Andhra Pradesh	Dudhia (Resistance to Powdery mildew), Banarasi

Propagation

Ber is commercially propagated by budding. In budding ring budding is commonly practice but it does require scion and root stock of same thickness which is generally not available. Therefore T-budding has been found successful. The other budding methods like patch, modified, ring, forkurt have been found most successful.

Budding: It means rising root stock seedling at permanent place and budding over them at appropriate time. The seeds of *Zyzyphus rotundifolia* are in-situ during rainy season. The germination takes place more than one month and seedling are allow to grow in next April. Seedlings are headed back to (15-20 cm) to regenerate new growth for budding. Budding is done in July under north Indian condition in June.

Rising plant in nursery: In-situ method budding can only practice where bud is locally available which is sometime difficult in villages under such circumstances to raise wide plant from the nursery. Following are the method of raising budded plant in the nursery.

1. In-situ seeds are sown in well prepared pit at appropriate distance in March to April such plants if properly care are ready for budding in July to September. This practice is highly suitable for arid condition and at the same time if budding are 1st prepared in nursery than lifted and transplanted they undergo great losses due to mortality for their inability to stand transplantation due to root damage.
2. The preparation of buddlings seedling rise in the nursery at a distance of 30×30 cm and budding is done in the month of September-October. If seedling raise in polythene bag in April to May followed by transplanting in field in July to August for in-situ budding. 90 days old seedlings are budded when they are still in bag and they transplanted in August in field. In case the material is to be transported to a distant place to overcome the problem of transportation.

Budding time: Budding time has been quite varied from place to place. But June to September appears to be quite successful. Stooling and air layering have also been reported to be successful with the use of auxin like IBA and NAA.

Planting

Beginning of monsoon is best time for planting in North India where as the period of planting can be extended up to November in South India. Distance of planting depend upon the climate and weather of the crop which is grown In rain fed condition distance will vary from 6x6m to 8x8 m.

Training and Pruning

Principally pruning is determined by fruiting behavior of a crop. In ber the fruits are born in the axils of the leaves on young shoot on the current season. Therefore pruning is required every year to induce maximum number of new healthy shoot which bear good quality fruit. Pruning consists of heading back of previous season growth. It should be done during the hot and dry season when the tree sheds its leaves and become dormant. The time of pruning in North India varies from 1st fortnight of May to 1st fortnight of June. The intensity of pruning depends upon the growth of the plant. In extreme arid pruning should not be done more than 15 per cent of the current season growth.

Top Working

A large number of wild ber tree are seen growing on regulated land road side and from border. These trees bear very small fruit with very poor quality and do not fetch any income. These tree can be made productive by top working with better variety for this wild tree are headed, retain a clean trunk up to one or two meter in height. It is however better to retain two or three main branches above 1-1.5 m height on the tree trunk. These branches are beheaded to 30 cm each. This can be done from March to June. However some scientist recommended that beheading should be done when the tree are dormant. After beheading the tree-sprout to give beddable shoot in 15-30 days depending on the age of the tree and the season when beheaded. Budding on the new shoot can be done during April to July in western Rajasthan by T-budding. Some scientist recommended budding in July-August therefore beheading of tree should be done and such a time beddable shoot develop during this period when budding is to be done when bud stick of the improved cultivar are also available. On each branch only one shoot should be budded with the scion from improved cultivar and the other should be removed. Such top work tree start bearing in the same year.

Nutritional Management

It is an important factor in dryland which ultimately affects the production. Nutritional requirement will vary from the location. Therefore the correct quantity to be applied can be determined after analysis of soil nutrient. The followings are some general recommendation.

Age of the Plant in Year	FYM (Kg)	Nitrogen (gm)	Phosphorus (gm)	Potassium (gm)
1	30	100	100	50
2	60	200	200	50
3	90	300	300	100
4	120	400	400	100
5	150	500	500	100

Full dose of Phosphorus and potash should be applied with the advent of rain along with the half quantity of Nitrogen. The 2nd dose of nitrogen should be given one month after the fruit set. It will desirable to apply the total quantity of fertilizer and manure with Monsoon and rain fed orchard.

Irrigation

Generally once established irrigation is seldom practiced in ber orchard. However the supplementary irrigation generally benefits trees and higher production is obtained. From June to October no irrigation is required being rainy period. In summer during April to June the plants being dormant require no irrigation but during post Monsoon period during October to March when crop is on if irrigation is practiced it benefits through higher return. Interval of irrigation is determined by many factors young plant shall require watering at an interval of 4-5 days during their 1st two month after planting in absence of rain.

Flowering and Fruiting

Flowering in August in North India but fruit set does not take place till middle of September. Flowering duration is from August or September to November with full bloom in October. Fruit ripening start from November, December to early January in Andhra Pradesh Maharashtra and Gujarat. In case of Uttar Pradesh, Haryana and Punjab fruit maturity is delayed up to March. It is extreme of temperature prevalent in these states.

Insect Pest and Diseases

Insect Pest

1. **Fruit fly (*Carpomiya vesuviana*):** This is the most dangerous insect of ber crop infestation start in September when the fruits are very small and continue up to the harvest. The adult female lays egg singly by inserting it ovipositor in the developing fruit. After two to five days larva hatches out and start feeding on pulp making galleries for 9-20 days during its life. The infested fruit become deform and their growth is checked and sometime they rot and become unfit for consumption.

   Control: These pastes are quit tedious however the following approach is quite effective.

a. Cultivation of orchard soil during summer month shall expose the pupae to death through damage are bird.

b. Application of 0.1 per cent aldrin or di-aldrin or themate @ 25 kg per hectare during the month of April may to kill pupa.

c. The dropped infested fruit should be collected and buried deep in soil to eliminate larvi in such fruits.

d. Spray of Rogar 30 EC or a mixture of (100 ml Mailathyan 50 EC) or 1 kg Jaggery in 100 liter water during fruiting season or very effective control.

e. Katha, tikdi, dudhia and mehrun are considered to be resistance varieties to fruit fly.

2. **Bark eating caterpillar (*Inderbela quadrinotata*):** This is the most dangerous insect of Ber crop. In this case the caterpillar remains concealed in the stem during day time and comes out to eat bark in night and make galleries on the stem and branches.

   Control: Control of this insect is very easy and chef. Galleries should be remove and visibility are exposed hole should be painted with a mixture of 10 ml, Monocrotophas 40 EC. or 40 ml of Triclorophas 50 EC or 40 gm of Mailathyan 50 wp in 10 liter of water during September to October. A few drop of Kerosene oil or Petrol is inserted in the hole also control the insect.

3. **Hairy caterpillar (*Euproctis fraterna*):** The caterpillar cause heavy loss of young leaves and tender fruits. This caterpillar can be controlled by application by (Carborile 0.1 per cent or Cyper methrin 0.1 per cent)

Diseases

1. **Powdery mildew (*Oidium spp.*):** This disease can be seen in the form of white powdery growth on the flowers, fruits and rarely leaves also. It does appear on fruits during October-November when fruit set is over. This disease can be control by spray of 0.2 per cent kerathene or application of supher containing fungicide like Sulphex or Bavastin.

2. **Black spot of leaves (*Isariopsis spp.*):** This disease has principally being observed in Haryana and Punjab. The blackish spot appear on the lower side of the leaf giving black sooty appearance. This can be control by spring of (0.3 per cent Bliatix or 2:50:50 Bordex mixture 0.2 per cent).

3. **Alterneria leaf spot (*Alterneria spp.):*** This disease appears in January-February with the formation of small inregualar dard gray spot on the ventral side of the leaf. This can controlled by application by (Difoltan.02 per cent or Dithen Z-78 0.2 per cent).

4. **Fruit rot (*Phoma spp., Colletotrichum spp., Alterneria spp., etc.*):** This is the post harvest disease and controlled by application of (02 per cent Dithen Z-78).

Harvesting and Yield

Fruit is harvested from November to March in different part of the country which is earlier in South and late in North India. The average yield of different variety at their prime age of bearing varies from 80-200 Kg per tree in dry area yield is lower which is about 50-80 Kg per tree.

Chapter 16

Citrus

Botanical name:	*Citrus* spp.
Family:	Rutaceae
Chromosome no.:	2n = 18
Origin:	Asia

Introduction

Citrus is the leading tree fruit crop in the world. In India, citrus ranks second in area sharing 12.8 per cent of the total area under fruit crops with 9.6 per cent share in production. Under citrus group, mandarins are the most important, occupying 50 per cent of the total area under citrus followed by sweet orange and limes. Citrus fruits are grown mainly in Maharashtra, Andhra Pradesh, Punjab, Karnataka and North-Eastern region. South East Asia, tropical and sub-tropical regions of Indio-china. Major citrus growing countries are USA, Spain, India, Italy, Japan, Argentina, Mexico, Brazil, Morocco, Algeria, Greece, South Africa, Australia, Israel, Egypt, Jamaica, *etc.*

Taxonomy and Morphology

Citrus fruits are a group of fruits comprising of mandarins, sweet oranges, grape fruit, limes and lemons *etc.*, Citrus fruits have a prominent place among popular, extensively grown tropical and subtropical fruits. They are small, spiny shrubs or trees with alternate, usually evergreen, leaves, which are shiny and leathery and dotted with oil glands. The stems are mostly winged and jointed with the leaves and there is usually a spine on the twigs at the attachment of each stem. Most of citrus species fruits are polyembryonic nature like *Citrus reticulate* (Mandarin) *C. unshiu,* (Satsuma mandarin) *C. deliciosa,* (Willow leaf mandarin), *C. nobilis,* (King mandarin) *C. limon*, (Lemon), *C. jambheri*,(Rough lemon), *C. aurantifolia*, (Acid lime), *C. limotoides* (Sweet lime), *C. sinensis* (Sweet orange), *C. aurantium* (Sour orange), *C. paradisica* (Grape fruit) and monoembryonic nature are *C. grandis* (Pummelo).

Composition and Uses

The composition of the citrus (although it varies according to the variety) is made up of approximately 75-90 per cent water, 6-9 per cent sugars and the rest includes pectins, minerals, dietary fibre and essential oils. Citrus fruits and juice contain various components that include calcium, potassium, phosphorous, vitamin C, vitamins B, dietary fiber, *etc.*, which are beneficial for the human body, also, there are various active compounds extracted from citrus used as medicines for heart diseases, and in treatments of hypertension. Citrus products have different healthy properties such as anticancer, inflammation, antiviral, antibacterial, and antifungal activity.

However, citrus fruit is still being used for different purposes like juice, jam, jelly, squash, pickles, cake, candies, marmalades, *etc.* Most citrus waste materials are currently used as animal feed.

Climate and Soil

Citrus belonging to the subtropical climate but thrive well in frost free area the climatic factors like light and temperature are much important in citrus cultivation. For good quality fruits citrus require dry semi arid climates. It can tolerate occasional light frosts. Optimum temperature range is 16-20^{0}C within a range of 17-40^{0}C and annual rainfall of 500-775mm is optimum. Mandarins grow well in frost free tropical and sub-tropical regions with an elevation of 370-1500m, average temperatures of 10-35 ^{0}C and an annual rainfall of 1000-12000 mm. Hot winds and severe heat during flowering and fruit set result in heavy loss of yields. Kinnow cultivar requires light chilling for good yields. Acid lime grows well in areas which are free from frost, strong winds and are warm and moderately humid. Can grow up to 1000m elevation with annual rainfall not exceeding 750 mm. Sweet limes can tolerate frost better than acid lime. Rangpur lime grows well under areas with 20-30^{0}C. Lemons can be grown in both humid and semi humid areas up to 1200m elevation and tolerate frost better than acid lime.

Citrus can be grown in a wide range of soil but thrive well in deep, loose, well aerated sandy loam soil. The ideal pH of citrus cultivation is 5.5-7.5. Mandarins grow in a wide variety of well drained soils but light loamy soils are best. It is grown in soil pH range of 4.5-9.0 depending upon region and type of soil, but 6.0-8.0 is an ideal pH. Lime grows well in deep (2-2.5m) well drained soils rich in organic matter. Optimum pH is 6.5-7.0 it cannot grow in water logged soils, alkaline soils with high lime content. Lemons grow better in shallow soils also if well drained.

Species

The classification of citrus was done by two famous eminent taxonomist, W.T.

Swingle (USA) and T. Tanaka (Japan). Swingle (1948) recognized only 16 species under the genus *Citrus* whereas Tanaka (1954) described as many as 144 species. He failed to cover many forms of horticultural importance and many species of Japanese, Chinese and Indian origins have been denied. Tanaka's classification although considered more comprehensive and detailed, contained excessive number of species, some of them being of doubtful validity. The mandarin group alone, he described 35 species, resulting into much confusion and obviously to lesser practical utility. Tanaka divided the genus *Citrus* into two subgenera *viz*; Archicitrus having 98 species and Metacitrus with 46 species.

Swingle divided the genus *Citrus* into two subgenera *viz*; *Eucitrus* having 10 species and *Papeda* having 6 species. Contrary to Swingle opinion, citrus forms of hybrids and certain cultivars by Tanaka were questionable and might be avoided. It is accepted that the characters employed for identifying a valid species of citrus should be free from the environmental influence.

Important Varieties

- ☆ **Orange:** Blood red, Jaffa, Hamlin, Pineapple (exotic), Mosambi, Sathgudi, Maltablood Red, Washington Navel Orange, Shamouti, Succari, Dobla Fina, Mudkhed (bud mutant of Nagpur mandarin).
- ☆ **Mandarin:** Coorg, Khasi, Nagpur, Kinnow (King mandarin X William leaf), Nagpur Santara, Satsuma Mandarins, *etc.*
- ☆ **Kagzi lime**: Pramalini, Vikram, Saisarbat- supposed to be canker free and prolific bearers.
- ☆ **Acid lime**: Chakradhar, Seedless strain PKM-1 heavy yielder. Selection 49: tolerant to canker, tristiza and leaf miner. It is also a prolific bearer.
- ☆ **Sweet lime** Mithachikna (Thin rind), Mithotra (Thick rind) and Sharbati.
- ☆ **Lemons**: Eureka Lisbon, Villafranka, Lucknow Seedless (Assam lemon), Nepali Round, Kagzikalan, Plant Lemon 1, (Selection of Kagzi Kalan tolerant to canker tristiza and die-back), Baramasi, *etc.*
- ☆ **Pummelo:** Kaopan, Buntan, White fleshed fruit and Red fleshed fruit.
- ☆ **Grape fruit:** Poser, Ruby, Marsh, Duncan Seedless, Thompson, Red blush, Triumph, Sharanpur Special.

Propagation

Citrus can be commercially propagated by seed. In asexual propagation T-budding is more common in sweet orange, mandarin, pummelo, Rangpur lime, rough lemon *etc.* and the best time of budding is January-March or September-October. In lemon and sweet lime air layering is commercially practiced. Jatti Khatti (*C. jambheri* x *C. karna*), Troyer Citrange, Kharna Khatta are popular root stocks for blood red in Punjab. Rangpur lime is popular rootstock for mandarins and sweet oranges. Sweet limes can be propagated by sub-terminal stem cuttings or layering's or budding. Lemons by budding on trifoliate orange or Rangpur lime (*Citrus limonica*).

Planting

Citrus planting should be done at an spacing of 5-8m. Its depend upon the species and rootstock like trifoliate orange. For acid lime spacing should be maintain 5x5m the best time for planting is rainy season and February-March. Befor planting the plant pits of 60x60x60 cm are prepaired and filled with the soil FYM and carbofuron. Mandarin fruit spacing is 4-6m. Kinnow on Troyer Citrange can be planted at 1.8mx1.8m spacing (3000 plants/ha). Lime and lemon can be planted before or after rains during the rainy season depending upon amount of rainfall. Spacing should be 4-6m for limes, and 6-8m for lemons. Pits can be of 1mx1mx1m and filled with mixture of soil and manure.

Crop Regulation (Bahar Treatment)

In subtropics, tree flowers in spring while in tropics, it flowers in three distinct phases *i.e.* Ambe bahar, Mrig bahhar and Hasth bahar. Spring flowers (*Ambe bahar*) give fruits in summer when the demand is maximum but quality is poor. Therefore, efforts should be directed to avail of rainy season flowers (*Mrig bahar*) so that fruiting period coincides with maximum water availability time and the crop is taken without much irrigation. For '*Mrig bahar*' treatment, withholding of water from December to May results in sufficient suppression of growth. Trees usually shed leaves by March end and remain dormant up to May. Then the field is ploughed, manure and fertilizers are applied and first light irrigation is given in the middle of May, and thereafter one or two light irrigations are given until rain sets in. Trees put up new growth '*Hasth bahar*', flower and fruit with crop availability in October to November. This is the most advisable approach in arid and semi-arid regions with limited resources. Flower regulation is done by withholding water a month or two before flowering till the plants wither and drop some of their leaves. Then they are manured and irrigated which results in profuse flowering.

Nutritional Management

The bearing tree should be given the fertilizer dose at the ratio of 550gm: 370gm: 550gm NPK/plant/year, depending upon its performance in sweet orange. Graded dose of fertilizers can be applied from 1st year to 10th year. Fertilizers should be applied in a ring 30-40 cm wide just below the canopy of the tree at a distance of at least 1-2 m from the trunk. For improving the fruit quality the application of 0.2 per cent magnasium sulphate and 0.1 per cent boric acid in the month of April. A fertilizer dose of 450 g of nitrogen, 450 g phosphorous and 900 g of potassium per plant in two split doses 1st dose in may and 2nd after harvest may be given from the beginning. As mandarin also shows micronutrient deficiencies, a composite spray of the following is recommended. Fully grown lime may be given 5 kg FYM, 900g N, 250g P and 500g K per plant per year. Total P, half of N and total K after rains and remaining dose should be given after flowering.

Irrigation

Irrigation requirements depend upon soil and weather conditions. Water should never come in direct contact with the trunk of the tree, for this reason double ring or check bund method is best suited for this crop. Presently drip irrigation is becoming popular which helps in saving of irrigation water. Stopping irrigation 1 or 2 months prior to flowering is beneficial to the crop, till the tree withers and drops half of its leaves. In summer crop should be irrigated at 8-10 days interval and in winter 12-15 days interval. Drip irrigation was found to be more beneficial. Water stress during growth, flowering, fruit set and development should be avoided. In south India, mandarins are grown under rainfed conditions in heavy rainfall areas. Limes require more water; irrigation may be given if top 25cm soil becomes dry. Check bund system is best as it will not allow the contact of water with the stem and drip irrigation improves yield.

Flowering and Fruiting

The main blooming period of citrus tree in subtropical climate is the spring from February-April, although seasonal condition may cause citrus to bloom at various times. Sathgudi orange in South India flowers twice in a year from December-April and September-December while Coorg mandarin have to distinct flowering period one during March-April (Main season and the other during September-October (Off season). In central and western India orange flowers thrice in a year at interval of four months *i.e.* June, October, December and February. Under tropical condition limes, lemons and citron flower almost continuously throughout the year, though spring is the main blooming period. In Assam most of the lemons flowers blooming throughout the year.

Insect Pest and Diseases

Insect Pest

1. **Citrus psylla** (*Diaphorina citri*)- This is the sucking type pest and they suck the sap from tender part. It can be control by the application of monocrotophos 0.05 per cent or phosphamedan 0.2 per cent or diamethoate 0.05 per cent.
2. **Lemon butterfly** (*Papilio demolius*)- This desiese is most dangerous in nursery stage. Insect eats the leaves and in severe condition plants are almost leafless. Insect infestation is more in the month of April-May and August-October. It can be control by application of carbaryl 0.2 per cent or a biotic agent like *Bacillus thuringiensis* 0.05 per cent solution.
3. **Leaf minor** (*Phyllocnistis citrella*)- This disease is most dangerous in nursery stage as well as young plant. In severe condition insect eat the skin of the leaves. It can be control by 0.05 per cent solution of monocrotophos.
4. **Aphid** (*Toxoptera citricidus, Toxoptera aurintii*)- In India beside two species of aphid the *Aphis gossypii* and *Myzus persicae* are the equally

important. This is the sucking type pest and infestation is more in the month of December-March this insect is suck the sap from leaves and inflorescence. This is also the transmitting agent of tristeza virus. It can be control by 0.05 per cent solution of nicotin sulphate or 1 per cent solution of neem oil.

Diseases

1. **Tristeza-** It is transmitted by aphid and specific symptoms of tristeza is pitting of inner phase of the bark or trunk severe pitting develop in lime species Citron while sweet orange, mandarin, rough lemon and sweet lime the stem pitting is usually absent. Affected tree develops various deficiency symptoms like leaf falling, root decay, dieback and ultimately plant die.

   Control: It can be control by the use of tolerant rootstock like rough lemon sweet orange, rangpur lime, trifoliate orange, by the use of bud wood free from viruses. It can also be control by checking the aphid population through insecticide.

2. **Greening:** It is caused by mycoplasma. The affected plant show chlorosis of leaves similar to zinc deficiency, thickening of leaf blade, saltaning of internodes, off season flowering, leaf shading and dieback. The disease spread through infected bud wood. Insect vector *i.e.* known as Citrus psylla (*Diaphorina citri*) is the main transmitting agent.

   Control: It can be control by the removal of the infected tree and controlling the insect vector by the use of insecticide.

3. **Citrus canker** (*Xanthomonas citri*)- It is wide spread in India. It is a serious problem in kagzi lime. The disease affects all plant part like leaf, twig, fruits *etc.* Canker lesion appears as minute water soaked roundish spot which in large slightly and turn brownish corcky. It can be control by the pruning of the affected part and spraying of 1 per cent solution of Bordeaux mixture on emerging flush or application of streptomycin sulphate 500ppm.

4. **Phytopthora rot-** This is also known as fruit rot, collar rot, crown rot, root rot and phytopthora gummosis. There are three species of phytopthora *viz.* Phytopthora parasitica, phytopthora citropthora, phytopthora palmivora has been reported as the casual organism of this disease. Fungus is soil born it's growing well under high moisture condition. It can be control by using the resistance root stock and provide good drainage facility.

Physiological Disorders

1. **Fruit drop:** This is serious problem of fruit drops in India where the two types of fruit drops occur. The first drop is June drop which is more common in May-June and second drop is pre-harvest drop which is more common in October-March. The application of 2, 4-D 8ppm in the month of April can control the fruit drop. The spray of NAA 80ppm in the month of September- first week of October also minimizes the drop. Fruit drops may be causes by climatic factors, improper water management, lack of nutrition *etc.*
2. **Granulation:** Mosambi, Hamlin and blood red cultivars are mostly affected by this disorder. It is drying up, becoming hard with greyish colour and enlargement of the juice vesicles with increase in pectin, lignin, *etc.* resulting in reduction of the juice content. This occurs more in young vigorous trees than old trees. The granulation is mainly affected by climate (humid), cultivars, rootstocks, mineral nutrition, enzymes and plant growth regulators and crop load.

   It can be control by spraying lime, ZnSo4, Bordeaux mixture or $ZnSO_4$ + $CuSO_4$ (0.5 per cent), 200-500 ppm lead arsenate. Early picking and reduced irrigations is also helpful.
3. **Fruit cracking:** This disorder is more common in lemon but sometime it is in malta also. It is mainly due to moisture fluctuation in soil (Irregular interval of irrigation). It can be minimize by application of potasium sulphate @ 4g/liter of water in the month of April, May and June.
4. **Citrus decline:** It occurs in mandarin fruit crop. It may advance in badly maintained orchards in ill drained soils and also due to malnutrition and insects and pests. Tristeza a very serious viral disease causing various deficiency symptoms in leaves like leaf falling, root decay, dieback of shoots and finally the death of the tree. Trees normally produce heavy crops before dying.

Though, appropriate control measures are not available as a means of rejuvenating the declining plants, the following measures may be followed.

1. Removing all dead wood before new growth starts and spraying Bordeaux mixture and covering all cut surfaces with Bordeaux paste.
2. Regularly following the recommended plant protection and spraying 0.3 per cent zinc sulphate and Bordeaux mixture whenever new growth appears.
3. New plantations should be with plants grafted on Rangpur lime which is reported to be highly tolerant to tristeza.

Harvesting and Yield

Being non-climacteric should be harvested only after full maturity of the fruits. Sweet orange takes 9-12 months for maturity. Harvesting seasons are December-February in North India, October-March in South India, November-January (*ambe bahar*), March-May (*mrig bahar*) in central and western India. Mandarin starts bearing from 4 years but commercially yields can be obtained from 10-12 years. Harvesting season starts from August and extends up to April depending upon the locality and the variety grown. Harvesting season extends from January-September depending upon the locality for acid lime and May for lemons and June-August for Rangpur limes.

Yield of sweet orange varies from 500-2000 fruits/tree depending upon the variety, agro-climatic conditions and age of the tree. Mandarin yield may vary from 500-1000 fruits/plant. Lime yields 2000-5000 fruits/plant.

Chapter

Chironji

Botanical name: *Buchnania lanzan* L.

Family: Anacardiaceae

Chromosome no.: 2n = 22

Origin: South East Asia/China

Introduction

Chironji is locally called as, char, achar, charoli, and chawar in different parts of the country. Information regarding area and production of this fruit in India is not available because it is not grown on plantation scale. They can be seen growing in forest areas. The production in India is mainly concentrated in the drier states and the produce is collected by the villagers. Its cultivation may spread to arid and semi-arid areas of Bundelkhand, resource poor areas and wastelands where other crops cannot be grown.

A tree of the dry region is found in north, west and central india. Plants may be seen in forest area of Uttar Pradesh, Madhya Pradesh, Chattishgarh, Maharastra, Bihar, Jharkhand, Orissa, Andhra Pradesh and Gujrat. The tree is having 13-18 meter height with straight trunk, young branches clothed with silky hairs.

Taxonomy and Morphology

Leaves thickly coriaceous, broadly oblong, obtuse, sometimes emarginate, glabrescent above, more or less villous beneath, articulately veined, the nerves and veins impressed on the upper surface, base rounded, main nerves 10-20 pairs, petioles long (about 1.2cm). Flowers small, sessile, greenish white, in terminal and axillary pyramidal ferrugineo-pilose panicles which are shorter than the leaves, bracts small, caducous. Calyx lobes short, broadly ovate, ciliate. Petals long, ovate-oblong, subacute, disk fleshy. Stamens 10, a little shorter than the petals, filaments flattened, anthers about as long as the filaments. Ovaries 1 perfect, conical, villous, the other 4 reduced to cylindrical filaments. Drupes obliquely lentiform in the long diameter, black, stone hard, 2 valved.

Climate and Soil

It prefers tropical and subtropical climate and can withstand drought admirably. Chironji is very hardy plant and thrives well on rocky, gravelly red soils and also on saline and sodic soils. Though it is a very hardy tree but plants do not survive under waterlogged conditions. It can grow even in pockets of soil between crevices of barren rock. Trees are grown on degraded rocky area including salt-affected soils. However, for its better growth and productivity, well-drained deep loam soil is ideal.

Important Varieties

Thar Priya: It has been identified as variety at Central Horticultural Experiment Station, Godhra. The selection out performed in respect of flowering pattern, fruiting and fruit quality attributes. The tree has spreading growth habit, thick trunk, evergreen, dense foliage and drooping branches. It is regular bearer and flowers in the second week of February. Peak period of fruit set is fourth week of February. It ripens in third week of April. Fruit pulp has total soluble solids 23.50 ^{o}Brix, total sugar 15.61 per cent, vitamin C 46.35 mg/100g. Stone weight 0.41 g, kernel weight 0.09 g and kernel protein 30.00 per cent. Fruit yield per plant is 13.20 kg during 9th year of orchard life under rainfed conditions of hot semi-arid ecosystem.

Propagation

- ☆ **Seed propagation and raising of seedlings**: Chironji plants are generally propagated by seed giving a long gestation period (15-20 years) and large variability. per cent germination in freshly extracted seeds is poor because of hard seed coat on the kernels. Seed germination of 83.00 per cent within 18 days could be achieved with satisfactory seedling growth by mechanically damaging the stony endocarp before sowing in the month of June, plant survival of 93-96 per cent with satisfactory plant growth was obtained when one year old seedlings were planted in 60x60x60 cm pit filled with red soil and 10 kg of well rotten farmyard manure coupled with proper mulching around the plants during winter and summer season and with or without thatching. Seed treatment with sulphuric acid (5-7 per cent) was also found very suitable for promotion of seed germination in chironji. Seeds can be sown on raised beds or in polythene bags about 30 cm deep during June-July and it germinates within 25-35 days. The seedlings become ready for grafting after one year from the date of sowing.

☆ **Soft wood grafting:** When one-year-old seedlings raised for rootstock commence putting on new growth and the leaves are of light green colour, these are ready for soft wood grafting. Shoots of 3-4 months, which have prominent apical bud, are taken as the scion material. Defoliation of such shoots is done 8-10 days before detaching them from the parent tree for grafting operation. At the time of removal of these shoots, the apical buds should remain intact. The top of the new growth of the stock is cut and the scion is fitted by cleft grafting or splice grafting. The union is tied with 200-gauge polythene strip. If the selection of scion and rootstock is proper, 60-70 per cent success may be obtained during the month of July-August. Singh and Singh (2014) suggested that softwood grafting in July may be adopted for multiplication of chironji with 66.66 per cent success.

Planting

The seedling plants should be planted 10m apart, whereas grafted ones 8m apart. The pits of 1m x 1m x 1m size are dug and filled with a mixture of top soil + 25 kg farmyard manure up to a level of 30 cm from the ground level. After onset of monsoon, soils are settled and planting is done in the center of the pit during July- August. Drenching of pits with insecticides are required if there is problem of termites.

Training and Pruning

Training is very essential to develop the framework of chironji plants. Plants are allowed to grow straight with the help of stakes. The framework should be developed by encouraging the growth of 4-6 well-spaced branches on the trunk at a height of 90-cm. from the ground level. Pruning is not performed in chironji plants except removal of dead, diseased and crossing branches.

Nutritional Management

A dose of 10 kg farmyard manure, 100g N, 50g P and 75g K/plant should be given to one-year-old plant. It should be increased every year in the same proportion up to the age of 10 years. Thus fully-grown up trees require 100 kg farmyard manure, 1 kg N, 0.5kg P and 0.75 kg K. Farmyard manure should be applied during July-August. Half dose of N and full dose of P and K should be applied a month of July and remaining dose of N may be applied by the end of August under rainfed conditions. The manure and mixture of fertilizer should be spread under the canopy of plants and incorporated in the soil.

Irrigation

Plant can survive under rainfed conditions. If the irrigation facilities are available, basin system of irrigation should be used for young plantations. For fully-grown up trees, strip or bed system should be followed. Irrigation should be given during summer season at the interval of 15 days. Proper moisture after fruit set is important for retention and development of fruits.

Intercropping

Intercropping is intended to maximize land and space use efficiency to generate supplemental income particularly during the initial unproductive phase of the orchard to protect the inter space from losses through weeds, erosion, impact of radiation, temperature, wind and water and enriching it by nitrogen fixing legume crops. Compatible crop combination is necessary with regard to species, cultivars, planting method and sequence. Peas, gram, lentil, black gram, cowpea, cluster been may be grown as inter crops in the orchard. Experimental evidences are, however, not available.

Flowering and Fruiting

In chironji, flowers appear in the month of January- February in different agro climatic zone of the country. The time taken for complete development of flower bud from its visible initiation to anthesis varies from 18 to 28 days. New plantation of grafted plants start flowering after 4th year. The cause of poor fruit set may be due to pollination problem or self-incompatibility. Seedling trees of higher age have better fruit set than the tree of lower age group.

Insect Pest and Diseases

1. **Hoppers (Sucking pest):** This is the most damaging insect during the flowering season. Both adults and the nymphs damage the crop. They suck the sap from tender shoots and panicles. The panicles wither away and the fruit set is adversely affected. They excrete honeydew, which develops a sooty mould on the leaves and panicles.

   Control: It can be controlled by spraying with dimethoate (0.03 per cent) or phosphomidon (0.05 per cent) once at the time of panicle emergence and then again at the fruit set stage.

2. **Mealy bugs:** These insects are characterized by a waxy coating over their bodies. Mealy bugs are found in large number on ventral surface of leaflets, base of leaf petioles, tender shoots and even fruits, and suck cell sap from different parts. The leaflets become chlorotic and fall down and in case of severe infestation; there may also be premature fruit fall.

   Control: Spraying with 0.03 per cent dimethoate or phosphomidon (0.05 per cent) is effective to control the pest.

Diseases

1. **Phytophthora gummosis**: The fungus affects the bark and can penetrate in a limited way in to the wood. In advanced stages, the bark cracks, shreds in lengthwise strips as it dries and gum starts exuding. In case of infection, if the disease can be detected before the bark covering more than half of

the circumference of a stem is damaged, scraping of the affected portions with a little extra healthy tissue without injuring the wood.

Control: Application of Bordeaux paste (1:2:16-20 water) on the scraped as well as healthy portion around the infected zone, may help in combating the disease and recovery of the plant.

2. **Powdery mildew:** During flowering season, in the beginning the appearance of grayish whitish powder on the flower buds, fruit lets and rachis of the panicles denotes the attack of the fungus.

   Control: Application of Bordeaux paste (1:2:16-20 water) on the scraped as well as healthy portion around the infected zone.

Harvesting and Yield

The fruit growth was faster initially and slowed down while reaching towards maturity and followed sigmoid growth curve, deep purple colour appeared on fruit surface of during peak period of ripening, ripening time varied from third week of April to second week of May. Fruit yield 13.20 kg per plant may be obtained during 9th year of orchard life under rainfed conditions of hot semi-arid ecosystem.

Chapter 18

Custard Apple

Botanical name:	*Annona squamosa* L
Family:	Annonaceae
Chromosome no.:	2n = 14
Origin:	Tropical America

Introduction

Custard apple is one of the most delicious fruit mainly consumed fresh. It is very hardy, tolerant to drought, salinity and saline irrigation water to certain extent. It grows very well even on shallow soils and leaving it to nature, its flowering coincides with maximum moisture availability period. It also sheds off leaves during stress periods to evade moisture loss from plant tissues through transpiration and thus a most appropriate fruit crop for arid region. Under tropical if water is available, it can be made to flower any time of the year to obtain off season fruits: Area under custard apple is reported to be more than 40,000 ha mainly in wild form in the states of Andhra Pradesh, Tamil Nadu, Orissa, Assam and Rajasthan. However, well organised orcharding of this crop can be seen in Saurashtra region of Gujarat where in water scarcity areas this crop is slowly replacing pomegranate, which comparatively demands more water due to longer reproductive phase, with annual gross income about Rs. 45,000/ha. This crop is known by varied names like Sitaphal, sugar apple, sweet sop and sharifa.

Taxonomy and Morphology

The genus *Annona* belongs to the family Annonaceae, contains more than seventy species of woody shrubs and trees of which five are commercially important like *A. squamosa* (Custard apple), *A. cherimola* (cherimoya), *A. reticulata* (Bullock's heart), *A. glabra* (Pond apple) and *A. muricata* (sour sop), the biggest of all these, *Annona squamosa* is probably the most common in cultivation in India. The plants of this species are shrubs or small trees with fairly stout and smooth trunk reaching to a height of 5 to 6 m. The branches bear alternate two ranked oblong lanceolate to lanceolate leathery leaves with solitary or a small group of nodding flowers which are greenish yellow consisting of three fused sepals. Six petals in two whorls of which outer whorl are more prominent while inner is scaly. There are numerous stamens and fairly hairy carples in the centre. The fruit is large fleshy syncarp formed by the fusion of pistils and recepticle, subglobose, cordate or conical yellowish green and glaucous, to 3.5 inch across, tubercled, the carples falling apart at maturity.

Composition and Uses

The ripe fruits are rich in sugars. Its pulp contains about 73 per cent moisture, 0.8 to 1.5 per cent proteins and 14.5 per cent sugars, 0.3 per cent fat and 0.7 per cent mineral matter. It is quite rich in calcium (0.02 per cent). phosphorus (0.04 per cent) and iron (1.0 per cent). The calorific value is 105/100 g of pulp. Glucose is predominant sugar. Leaves have medicinal value due probably to annonine and the oil extracted from seeds could be used for soap and cake as animal's food. Basically this fruit is used for table purpose, however, in recent years it is being used in ice-cream and puddings.

Climate and Soil

It is a tropical plant, can be grown in mild sub- tropics also. It is sensitive to frost. It prefers dry climate during flowering but fruit set is aided by high humidity which begins with the onset of monsoons. Annual rainfall of 50 to 75 cm is optimum, though it can stand higher rainfall or drought. In tropics water stress puts this plant in quiescence while in sub-tropics low temperature and water stress in summer create these conditions which help this plant to stand drought better and makes it an ideal plant for arid horticulture.

Generally, shallow and poor classes of soil are preferred for custard apple cultivation, the kind of soil available in arids. It is tolerant to many soil conditions, except alkalinity, like presence of lime, chlorine, rockiness, shallowness, light and heavy with good drainage.

Important Varieties

Propagation of this fruit tree has generally been through seed and therefore, there are no standard varieties. Only two distinct groups are known like red fruited (*A. squamosa* var. rubra) and green fruited in which large variables are reported. Important varieties are Balanagar, Washingatan-107005, Red Sitafal, Island Gem, Mammoth and Arka Sahan which are commercially cultivated in India.

Propagation

Custard apple is generally grown from seeds. For arid conditions seedlings can be raised in polythene bags filled with garden soil and planted in field at the onset of monsoon when they attain the height of 20 to 25 cm. Seeds can also be sown directly in the field in well made and prepared pit in the beginning of monsoon and at appropriate height they can be softwood grafted when new growth emerges after one year.

Among vegetative methods of propagation budding gave 100 per cent success. However whip grafting is better than budding. *A. cherimola*, *A. squamosa* and *A. reticulata* can all be used as rootstocks.

Planting

Pits of 60x60x60 cm size are made in summer, refilled with soil + FYM (1:1). Planting distance is kept at 4.5 x 4.5 to 5 x 5 m. Planting is done in the beginning of monsoons to avail the advantage of available rain water.

Orchard Soil Management

In rainy season efforts should be made to conserve water in root zone by developing micro catchments. There can be many approaches for this conservation like preparation of circular basin of about 1 m diameter with 5 per cent inward slope and surface mulching with dry grass. Another approach is V-shaped micro-catchment on either side of the tree row with 5 per cent inward slope and mulching with grass on run on surface. After the harvest of crop in October- November the field can be left undisturbed to permit natural hardy grass cover to avoid erosion due to wind.

Nutritional Management

About nutrition of custard apple, research based data about the dose are not available. However, in Gujarat where well organised orcharding exists, application of 10 to 15 kg FYM, 2 to 3 baskets of river silt, 500 g of castor cake. Sufficient quantity of fertilizers to supply 50 g N, 50 g P and 50 g K per plant is in vogue. These are applied in the beginning of monsoons. It has observed that this plant responds better to organic matter. Research work done at Rahuri (Maharashtra) has indicated that application of 250 g of N, 125 g of P_2O_5, and 125 g of K_2O per cent gave maximum yield.

Irrigation

Custard apple does not require irrigation to produce fairly good crop provided effective water harvesting is affected in rhizosphere by developing micro-catchments in rainy season. However, in post-monsoon period 2 to 3 irrigations help in better quality fruits and higher production. Drip irrigation is the most efficient method to manage scarce water resource which can be created by harvesting runoff during monsoon. Cracking of fruits sometimes occur due to poor water relations, which can be overcome by regulated water management.

Insect Pest and Diseases

Insect Pests

Mealy bug (*Ferrica virgata*): This insect is very common in Gujarat and gets conglomerated at the stalk end or in the cracks between tubercles. It can be controlled by spraying Nuvacron or Phosphomidon at the rate of 10 ml/10 litres of water.

Diseases

Anthracnose (*Glomerella cingulata*): This can be controlled by three fortnightly sprays of Benlate 0.05 per cent from the beginning of initiation of symptoms or Dithane M-45 (0.02 per cent) or Bavistin (0.05 per cent).

Leaf spot (*Cercospora anonae*): This can be controlled by spraying copper fungicides at 0.1 per cent concentration especially Bordeaux mixture is most effective.

Harvesting and Yield

It takes about three years for plants to come into bearing and five years for developing full canopy. In subtropics flowering period extends from spring (February to March) to rainy season while in tropics it is mainly governed by the availability of water, therefore, in natural conditions during rainy season. An average plant shall bear about 2000 flowers and the fruit set is about 2 to 3. per cent. Dichogamy exists, therefore, benefitted by hand pollination. Fruit set is better when humidity is high. Set can be improved by spraying GA_3 50 ppm at blooming stage. A five year old healthy tree give yield about 50 fruits/year, which can go up to 100 fruits/year. Optimum productive life is about 25 years.

It is a climacteric fruit and should be harvested mature which can be judged on the basis change of colour between tubercles and their falling apart. They should be harvested firm for distant marketing. Harvesting period extends from September to November depending on the flowering period. This can be varied also.

Chapter 19

Date Palm

Botanical name:	*Phoenix dactylifera*
Family:	Palmae
Chromosome no.:	2n = 36
Origin:	Persian Gulf and Iraq

Introduction

Area and distribution: Although it is the fact that date palm is originated near India and became naturalized in this country in very early time. There are no commercial plantation of high quality dates worth the name in the India union accept some seedling plantation of great value and potentiality in Kutch region of Gujarat state where 11-12 lacks seedlings palm are known to exist. Attempts are now being made to grow it in Rajasthan, Punjab and Haryana. Most important countries producing date palm Iraq, Saudi Arabia, Iran, Egypt, Algeria, South California and Pakistan with the world population around 98 million tree.

Taxonomy and Morphology

Date palm belongs to family Palmae, genous *Phonenix* has been divide in to 12 species beside *Phonenix decetylifera* the names of *Phonenix atlantica P. canariansis, P. reclinata,* and *P. sylvestris* also deserve mention. Date palm is a monocotyledonous plant it is diocious in nature inflorescence is spadix, elongated flowers small yellowish born singly male flowers with samples univated with a lobed cupule, petels ovtuse or rairly acute to acuminate, valvate stamen usually fix with female flowers with sapels three univated in cupule, petels three imbricate, carpels three separate only one usually maturing. Fruits are cylinderic or oblong ellipsoid single seeded berr, 1-3 inch long with thick sweet eduble fress and cylendar often pointed seed with single groove.

Composition and Uses

Dried (*Chhuhara*) as well as mature fresh date fruits are liked by masses. For inhabitants of Arabian region date is a staple carbohydrates food. Dates are rich source of sugar and menial content. Edible portion of fresh date containing 75 per cent sugar, 2-4 per cent protein, 2-3 per cent minerals. Among these menials iron is 7-9 ml/100g, Ca is 0.1-0.3 per cent, P is 0.05-0.1 per cent. Dried date which are preserving as Chhuhara containing 2-3 protein, 80-85 per cent carbohydrate, 1.8-2 per cent ash. Dates are favourite food along with milk. In addition, dates are used to commercially produce. Date juice, liquid sugar, nectar/honey, alcohol, beverage extract, halwa, vinegar, chocolate, chips, sauce, paste, pickles, jam and marmalade, animal feed.

Climate and Soil

Date palm is very exacting climatic which according to an Arab saying should grow with its feet in running water and its head in the fire of the sky. It has ability to produce fruits abundantly as extremely low humidity at list supply of ground water suffices makes the tree one of the chef producer of food in the desert belt of North Africa and South west Asia. To be more precise its successful cultivation requires. Along summer with high day as well as night temperature. A mild winter (up to a fall of 2.2 °C for short period without frost and through absences of rain at the time of flowering and fruit setting with low relative humidity and plenty of sunshine.

In Indian context in subtropics fruiting period in confined only to February-July where ripening period consider with rainfall and therefore fruits do not rich ripening stage and had to be harvested at earlier stage of fruit development. The date growing areas are classifying in to four zones on the basis of climate.

Zone 1- Covering district of Jaisalmer and Badmer of Rajasthan.

Zone 2- Covering district of Kachh of Gujarat.

Zone 3- Bikaner, Nagpur and Jodhpur district of Rajasthan.

Zone 4- Abohar (Punjab) Shri Ganga Nagar Churu (Rajasthan) Hisar (Haryana).

Date palm can tolerate soil containing more alkali or salt compare to the fruit plant and can grow in soil containing 3-4 per cent while alkali but to bear well the palm roots must be in a stratum with less than 1 per cent of the alkali salt. The sandy loam 2-3 m deep with good water holding capacity and drainage are the best.

Important Variety

Variety of date is numerous and is generally divided into three group according to moisture contain in the fruits *i.e.* soft, Semi dry and dry. The immature date content high proportion of sucrose, cane sugar. As ripening progress sucrose is converted to invert sugar or reducing.

1. Soft varieties have relatively high water content, little or no sucrose when fully ripe also called invert sugar date.

2. Dry date or semi dry date pauses a high sugar and low water content and retain more or less sucrose when fully ripe also referred as cane sugar date.

    - ☆ **Barhee:** it is a soft date introduce form Iraq and is amber on ripening heaving little astringency in Doka stage but ripens late. It is the heavy yielder variety ranging up to 150 kg/plant.
    - ☆ **Halawy/hillawy:** It is a soft date from Iraq is does not get damaged or gets little damage by occasional rain and high humidity, it ripens early and yield 100-125 kg/plant.
    - ☆ **Khadrawy:** It is a soft date of Iraq, ripens early yield is 50-60kg/ plant.
    - ☆ **Madjool:** It is a soft date from morocco, facing extinction due to beyond disease to which it is highly susceptible. Fruits have been slightly damage due to occasional rains and high humidity, yield is 75-100 km/plant.
    - ☆ **Dayari:** It is a semi date from Iraq it does not get damaged by humid weather when cured attains a dark reddish brown color. A mid season variety and yielded 75-100 kg/plant.
    - ☆ **Degletnoor:** It is a semi dry date from Algeria it is a leading variety in USA. The yield is 100-150 kg. per plant.
    - ☆ **Thoory:** It is the dry date from Algeria it is only slightly damage by high humidity. Yield is 100-120 kg/plant.
    - ☆ **Zahidi:** It is a semi dry date from Iraq. Less tolerant to rain and high humidity. Ripened in mid season, yield is 100-150 kg/plant

Propagation

Date may be propagation by seed or offshoots (sucker) when grown from seed the population shall be half male and half female (1:1) and no to seedling shall be a like and a few of these are likely to produce fruits of good quality. However when a seedling palm appears outstanding in any way it can be propagated by offshoots. Which will always reproduce the parent type. Which develop from the auxiliary bud on the trunk and offshoot of three year or more obtain to 10-25 kg weight is preferred. Offshoots must have good root system. For quick propagation date nursery seed to be establish by planting offshoots of promising variety at a distance of 1.5 x 1.5m or 2 x 2m. When the parent plant discontinue offshoot than planting of offshoot at a distance of 8 x 8m.

Planting

First of all field is cultivated and leveled. The pits are then dug 1m x 1m x 1m during summer as preparation for monsoon. Let the pits stay open for 15-20 days. Thereafter they are filled with well-rotten FYM (Farm Yard Manure) + soil to fill the pits. The time of planting depending upon irrigation facilities. Before planting the date offshoot field should be thoroughly plough and dig the pith the size of 1x1x1m at a distance of 8x8m. The best time for planting is rainy season (July to September).

Training and Pruning

Date palm grows straight without producing any branch therefore it does not require training or pruning of leaves at least up to 5 years of age. However well grown up bearing plantation require leaf pruning. Pruning is essential to maintain enough number of green leaves for good growth and yield of date palm farming. Scientists believe that the tree should have 75-100 leaves at the bearing stage. This means, no pruning is required for the first 4-5 years, as only up to 20 leaves are formed annually. It is important to maintain sufficient leaf bunch ratio to improve fruit quality.

Nutritional Management

Application of 1.36 kg nitrogen per tree has been considered necessary which can be supplied through any nitrogenous fertilizer or manure. Manures can be applied in the beginning of winter season and the fertilizer in March-April, and adult tree should be given 600 gm nitrogen, 100gm, phosphor 70 gm, and potassium per annum. A young tree should receive 0.5 kg nitrogen per year up to bearing age. At bearing age it should be increase to 1kg.

Irrigation

Light irrigation is required immediately after plantation. Date palm crop is resistant to drought and excessive irrigation. Usually in sandy soil the irrigation should be given 7-10 days interval during summer and 15-20 days interval during winter. The young plants are more demanding and requiring frequent irrigation for better survival. The peak growth and precious bearing. Under arid run of collection technique will have to be adopted to facilitate maximum water harvest in root zone.

Flowering and Fruiting

Under better management practices date palm start flowering at the age of 4-5 years or so, after planting of offshoots. The palm starts producing adequate yield at the age of 10 years, which increase up to 20 years; thereafter natural yield potential stabilizes and maintains its potential up to 50 years or so. Date palm is dioecious and generally speaking there are no secondary characteristic from which the sex of young tree before the first flowering can be inferred. Just 5-10 per cent of male plants are enough to pollinate 100 female plants. Pollination is done through hand or machine in commercial date palm farming. About 3 male flower strands are inserted in female flower strands. Pollination is an important cultivation practice in date palm cropping, as it influences size and quality of dates palm fruits and time of ripening of dates. The phenomena metaxenia are also found in date palm (it is effect of pollen grain on the testa and fruit wall). The fruit thinning is generally done 6-8 weeks after pollination. A fruit set of 50-60 per cent are sufficient to harvest economic yield. Under optimum management they start bearing at 4-5 year age.

One palm plant can produce 10-30 bunches. Date palm comes in to full bearing at the age of 10-12 year. The fruits develop and mature in a sigmoid curve through four stages are:

i. **Gandora (Kimri)**- Hard green
ii. **Doka (Khalal)**- Fully grown, hard, yellow or red
iii. **Dang (Rutab)**- Softening the tips of full fruit
iv. **Pind (Tamar)**- Fully ripened and loose much of its water, sugar content goes high to prevent fermentation.

Fruit Thinning

Fruit thinning is necessary to avoid shriveling of berry, damage to the rain and humidity delay in ripening, alternate bearing, reduce size and quality of the fruit. This can be accomplished through bunch removal, fruit thinning are strand thinning. Fruit thinning by removal of 1/3 central strand for higher yield and quality has been recommended.

Insect Pest and Diseases

Insect Pest

1. **Red weevil or Indian palm weevil** *(Rhynchophorus ferrugineus)*: The eggs are led on the soft tissue at the base of the leaf the larvae bore tunnel through the tissue of the stem near the top. These tunnels cause withering of the top, drying of leaf and sometime breakage of the top. It can be control by any systemic insecticide like: Rogar, Monocrotophas, Dimecron, Dimethoate or cypermethrin @ 2-5 ml/liter of water.
2. **Black palm beetal** (*Oryctes rhinoceros*): The adult beetal boar through closed leaf and give out fibrus debris. In serious incidence top leaves wither and drop. It can be control by application of Mailathiyan or aldrin dust.
3. **Scale insect** (*Parlatoria blanchardi*): It attacks both leaf and fruit. Productivity goes down and fruit are rendered unfit for marketing due to ugly appearance. It can be control by application of Mailathian @ 0.01 per cent.

Disease

Smut (*Graphiola phoenicis*): In this disease brown spot appear on both the surface of leaves and heavily infested leaves dry. With the increase in humidity in environment the incidence of disease also increases. It can be controlled by spraying of Bordex mixture 4:4:50.

Harvesting and Yield

A growing date palm fruits posse through four distinct phases which are name variably in different languages: The popular versions are:

1. 1st stage is "Kimri" by Arab and "Gandora" in Punjabi - In this stage fruit is developing and is green in color.
2. 2nd stage is "Khalal" by Arab and "Doka" in Punjabi - In this stage the fruit had attain full size and color change from green to yellow or red depending on the variety.
3. 3rd stage is Rutab by Arab and Dang in Punjabi- In this stage the tips of the fruit soften.
4. 4th stage is Tamer by Arab and Pind in Punjabi- In this stage fruits are fully ripe and dry enough to keep well without spoiling.

In India fruits are harvested at Doka stage to avoid spoilage due to rain and high humidity to advance Doka stage spray of 1000 ppm Ethephon at color break stage has been recommended in Gujarat.

In Gujarat under dry farming condition the average production is 40-50 kg/ plant. However under irrigated condition 200 kg fruits per plant.

Chapter 20

Fig

Botanical name:	*Ficus carica*
Family:	Moraceae
Chromosome no.:	2n = 26
Origin:	South Eastern Asia

Introduction

Fig is an important fruit crops intensively used in fresh and dried form. It is intensively cultivated in the country like Italy, Spin, Turkey, Greece, Portugal, Algeria and States of California in USA. In India the major growing areas are Pune in Maharashtra and shri Rangapattanam in Karnataka.

Taxonomy and Morphology

Fig is a monoecious crop and tree bearing fruits in two distinct sex forms *i.e.* I. Fruts with long style pistilate flowers only and II. monoecious with both short style pistilate and staminate flowers in same fruits known as Capri fig. Pistilate form have been classified into three type: i. The common/Adriatic/Edible fig ii. The sanpedro and iii. Smyrna type

The important characters are these tree forms are as follows:

1. **Common type/Adriatic:** This type of fig mostly grown in India tree of this type may or not produce a first crop or Breba crop born on wood of previous season growth. They do produce a more or less profuse Second crop or even a Third crop late in the season in the axils of leaf of current growth both the crop are produce by parthenocarpy.

2. **Sanpedro type:** The first crop is produce by parthenocarpy but second crop require pollination.
3. **Syrmna type:** Trees generally produce main or summer crop only although a few breba may rich maturity by parthenocarpy. Main crop is non parthenocarpic and need pollination and fertilization.
4. **Wild/Capri Fig/Goat fig:** This type of fig is generally used as pollinizer.
5. **Capri fig:** The tree of second sex form the monocious or capry fig produce at list three crops annually being inhibited by the fig insect blastophaga psenes. The first the memme crop is initiated in the late fall remain on the tree during the winter with the insect in larval stage and mature in spring. The second profichi is the most prolific crop appearing on the wood of previous season growth. It mature in early summer with the staminate flower producing pollen profusely, in nature the pollen become attached to bodies of the insect as they immerge and is then carried to respectile of either capry fig where the insect oviposit for short style flower or to other fig having long style flowers unsuitable for oviposition. This process of transfer of pollen is known as caprification which help in formation of seeds or achenes which have been classified as druplets.

Composition and Uses

The fresh fruits have 84 per cent pulp and 16 per cent skin. The calorie value is 269. 4gm protein, 200mg calcium, 4mg iron, 100 IU vitamin A and 0.1 ml thiamine per 100g of edible portion. The fruits may be eaten fresh or dry, preserved, candied and canned. It can also be made in to Jam. Fruits are valued for their laxative properties and have grate medicinal properties and are rich in protein digestive enzyme.

Climate and Soil

Fig is a deciduas plant but does not require chilling to break dormancy. It can be grown in tropics and sub-tropics both but does not do well in the low wet tropics and can be grown at higher alleviation and areas of low rain fall during flowering and fruiting. North India remains dormant during winter and puts up new growth with the advent of spring season. However in western area during monsoon (August-September) it become dormant and puts up new growth in October. For high quality fruits it should be slightly warm during fruiting. Irrigated arid and semi-arid region are considered better for fig.

Fig can be grown in wide variety of soil but light soil with low level of organic matter. It can tolerate salinity and water logging to certain extent. Medium heavy soils are better for its cultivation. Light soil is not suitable due to high demanding the water.

Important Varieties

The important cultivated varieties in India are Poona Fig, Black Ischia, Brown Turkey, Bangalore and Marseilles.

Propagation

Fig is commercially propagated by hard wood stem cutting. Cuttings are taken from 2-3 year old shoot with 1-2 cm thickness and short internodes. There are planted in December-February in North India and in July-November in Central India.

Planting

For planting of fig tree digging of pit at 1x1x1m in summer season and filled with mixture of soil organic matter compost and single super phosphate with neem cake at a distance of 6x6m or 8x8m. Planting should be done in monsoon period or in early spring. At the time of heavy rain fall planting should be avoided.

Training and Pruning

The young plant should be trained properly to give proper head and scaffold branches. It can be trained on single stem or multi stem, Pruning of fig is important. It generally two crops in a year the first born on previous year growth is breva and the second crop on current season growth and therefore pruning intensity and its type will depend on bearing habit and type of crop desired. Practice are pruning is greatly varied. In some region the tree are headed back to about two buds of previous year growth every year to keep them dwarf. This reduce the production but quality of fruit is better while in other region light pruning is practiced which give higher production but quality of fruits is poor. A balance should be struck. Notching is also practiced to forced new fruit bearing shoot from the lower o the branches. For this purpose a half cm wide notch of bark is remove above two buds on the middle portion of the branch in July. In north India pruning is done in December to a level of 3-4 buds on previous season growth. Pruning benefits by controlling plant height and providing high quality produce, therefore it should be regularly practice.

Nutritional Management

An five year old fig tree require 125 gm Nitrogen, 120 gm Phosphors and 100 gm Potassium the time of manuring shall vary with place and accordingly in north India is done in spring while in South and West India September is a preferred period. Manuring should be done under the drip and mixed in soil.

Irrigation

Fig is fairly drought tolerance but during summer when the fruits are developing and ripening irrigation twice a month is common throughout India. This gives large and sweet fruit. Excess of irrigation makes the fruit insipid. Sometime excess of irrigation during ripening causes cracking of fruits. Therefore judicious watering is desirable for high quality crops.

Flowering and Fruiting

The fig bears two crops in a year. The spring crop breba is born on the previous year shoot. In Europe the top bud of these stocks are cutoff before spring. This is shed to give early maturing and higher yield. The second crop is born on the current season growth and is the main crop. In North India the spring crop ripening in May is taken as main crop. In central and south India the fig crops in July to September and February to May. The later crop is sweeter and more valuable. Indian fig develops fruit without pollination. The sheaths in the fruits are only hollow shells. Tree may start fruiting in three year of age but commercial bearing starts from year on word and continue to bear up to 50 year of age. In arid region bearing in July to September is preferred to synchronized with the period of maximum moisture availability.

Insect Pest and Diseases

Insect Pest

1. **Stem borer (*Batocera Rufomaculata*):** The grubs bore into the main stem or branches and make zig-jeg tunnel in the wood resulting in slow depth of affected part and reduction in production capability. It can be controlled by insert the hole with cotton deep in kerosin oil or petrol.
2. **Leaf eating caterpillar (*Ocinaria varians*):** This is a wild silk moth harmful to fig. Caterpillar fig on leaves and causes damage. This can be control by dusting of Mailathiyan or spray of Metasistocks 0.05 per cent.
3. **Leaf roller (*Phydeos minor*)**: The adult lays egg on tender leaves. On hatching the caterpillar fig on epidermis of fig leaf during spring later they fold the leaves singly or one over another and feed within. Defoliation take place adversely affecting the production. This can be control by application of 0.05 per cent solution of Endosulphon, Finitrothian or monocrotophas

Diseases

1. **Leaf rust (*Cerotelium fici*):** This is the most serious disease of fig in India. Small rusty spot appear on the lower surface. It defoliate the leaves causes reduction in fruit growth which ultimately drop off. It can be control by application of Bordex mixture 5:5:50.
2. **Leaf spot (*Cylindrocladium scoparium*):** Browny angular spot becoming light ashy color later on appear on the leaves reducing their photosynthesis efficiency and production capacity. This disease is more common in Uttar Pradesh and can be control by application of 0.1 per cent Bavastin.

Physiological Disorders

1. **Fruit splitting:** When fruits are nearing maturity a sadden fluctuation in soil and atmospheric humidity caused splitting. This is aggravated by

rains during fruiting. The maintenance of high level of organic matter in the soil can control the situation.

2. **Sun burn:** In young tree due to exposal of sun light sun burning and cracking of stem and eruption of skin is common. Lime painting is more beneficial to minimize the sun burning.

Harvesting and Yield

Fruits of figs are harvested when mature, which can be judge on the basis of color development. In North India spring crop generally ripen in May while in Central and South India the fig bears in July to September and February to May. Sometime fruits are allowed to fall from the tree of their own and allow drying and then collected. A good harvest may give 300-500 fruits per tree which depend on the size of the tree and method of training, Sometime 180 fruits per tree are born by the small bushes.

Chapter 21

Guava

Botanical name:	*Psidium guajava*
Family:	Myrtaceae
Chromosome no.:	2n = 22
Origin:	Tropical America between Mexico and Peru

Introduction

Guava is one of the most popular fruits of India. It was first taken by Spaniards from its origin to across the Pacific to the Philippines and by the Portugese from West to India. According to Bruton, in the early years of seventeenth century it was found growing in India. Many people consider it indigenous to India due to its widespread existence and popularity with people. Some of the cultivars developed here are best in the world. It is cultivated throughout the country and considered to be one of the hardy fruits capable of growing under adverse conditions of soil and climate.

It is the fourth and fifth most important crop by production and area respectively. It occupies an area of 1,62,000 ha with a total production of 14,75,000 tonnes. About 50 per cent of this is confined to Uttar Pradesh followed by Bihar, Madhya Pradesh, Maharashtra, West Bengal and Gujarat. Allahabad region of Uttar Pradesh is worldwide known for excellent guava cultivation especially the 'Allahabad Safeda' cultivar. This fruit has great future in arid and semi-arid regions of the country.

Taxonomy and Morphology

Guava is a small tree with great branching and suckering habit. Spreading branches with smooth and scaly skin. New shoots are small, angular and hairy.

Leaves simple opposite in pairs, elliptical to oblong in shape. Flowers axillary, solitary or in 2 to 3 flowered cymes, white in colour; calyx entire in bud, splitting irregularly into 4 to 6 lobes, reflected, pubescent, persistent; petals 4 to 5, white reflected; stamens numerous, inserted in rows on disc, filaments white, anthers pale yellow; ovary 4 to 5 locular; style filiform; greenish yellow, long exserted above stamens; stigma capitate. Fruit is a berry surmounted by calyx lobes, varied shapes globose, ovoid or pyriform; colour pale green, to bright yellow or red; flesh white, yellow, pink or red. Seed numerous embedded in pulp, bony, reniform. Seeds remain viable for about a year and germinate within 2 to 3 weeks. Self and cross pollination occur.

Other related species are strawberry guava (*P. cattleianum*), *P. guaneense*, *P. microphyllum* and *P. freidrichsthalianum*. They also produce edible fruits. *Feijoa sellowiana* (Pineapple guava) is another close relative with commercial significance.

Composition and Uses

Guava fruit contains about 80 per cent water, 1.0 per cent protein, 0.5 per cent fat, 13 per cent carbohydrates and 5.5 per cent fibre. Total soluble solids vary from 6 to 10 per cent. It is one of the richest and cheapest source of vitamin C varies from 100 to 260 mg to sometimes 400 g per 100 g pulp and above depending on the climate. Winter season crop is rich in quality compared to summer and rainy season crops. Guava is also a fair source of vitamin A, iron, calcium and phosphorus.

Ripe juicy sweet types are mainly eaten fresh. The fruits after the removal of the seeds are made into preserves, jam, jelly, paste, nectar, and beverages. The greatest commercial use is jelly for which sour types are preferred in West as they make better jelly. However, in India sweet types are only used for jelly also. In some countries the leaves are medically used against diarrhoea and for dyeing and tanning.

Climate and Soil

Guavas can successfully grow in tropics and mild subtropics from sea level to 1500 m. In tropics with optimum temperature and humidity, it fruits throughout the year, however, in subtropics it has distinct seasons of flowering and fruiting. Optimum temperature range is 15 to 30°C. It is sensitive to frost and can tolerate high temperature to considerable extent (46°C). Severe winter delays maturity of fruits and size also remains smaller. It is sensitive to high rainfall which brings down the quality of fruit. However, adaptability range is quite wide. To be precise, in India the best guava grows in Allahabad region of U.P. which provides an indication that it requires distinct but mild winter for developing good fruit quality which is greatly influenced by the climate.

The guava tree has immense capacity to grow under wide variety of soils starting from heavy clay to light sandy and gravelly where most other fruits would not survive. The importance of guava lies in the fact that it is a hardy fruit which can grow under alkaline or poorly drained soil without any manuring or irrigation.

It can grow in soils with the pH from 4.5 to 7.5 and can tolerate temporary water logging. Guava tree fruits only on new growth, therefore, well drained fertile loamy soil is considered to be highly suitable. It is believed that soil may be fertile to any extent but it can't be said to be too fertile for guava. Therefore, this tree would show great response to irrigation, manuring and cultivation.

Important Varieties

Multiplication through seed used to be in vogue. Therefore, considerable variability exists. With the standardization of vegetative methods of propagation, they have been stabilized. About 92 varieties are known but 10 to 12 varieties are commercially popular and cultivated.

Basically guava varieties can be grouped into two (1) seed- less and (2) seeded. Among seedless 'Nagpur seedless' and 'Saharanpur seedless' are popular and they are triploids with deshaped fruits and very low productivity, commercially insignificant. Seeded varieties can again be subdivided on following bases:

a. **Based on shape:** 'Pear shape' and 'Karela'.
b. **Location:** Allahabad Safeda', 'Nasik', 'Dharwar', 'Dholka".
c. **Colour:** 'Chittidar', 'Red flesh", "Apple colour".
d. **Surface:** 'Dharidar'. 'Reshamdi'.

Among available variables 'Allahabad Safeda', 'Sardar', 'Behat 'Chittidar', 'Red flesh' and 'Apple colour' are in U.P. 'Nasik', 'Sardar', 'Dharidar', 'Dholka', are popular in Maharastra and Gujarat; while 'Dharwad', 'Bedana". Nasik' are main varieties of South India. From Andhra Pradesh some hybrids have been released 'Safed Jam' ('Allahabad Safeda' x 'Kohir') and 'KohirSafeda' ('Kohir' 'Allahabad Safeda").

- ☆ **Allahabad Safeda:** Most popular variety of U.P. Tree of medium height, medium branching; fruits medium to big, round, smooth surface, yellow colour; flesh soft, white sweet with good flavour; seeds many, big and hard.
- ☆ **Chittidar:** Identical to Allahabad Safeda' with many pinkish red dots of pin head size on the surface of the fruit. Flesh is white.
- ☆ **Sardar Guava (Lucknow-49);** It is a selection by Dr. G.S. Cheema in 1927 from seedlings of 'Allahabad Safeda' made at Ganesh- Khand Fruit Research Station, Pune. It has large tree with spreading and drooping branches, large fruits with rough skin. Fruits medium to big, oval shaped, rough surface, and yellow coloured; flesh soft, white, sweet with good flavour. Seeds less, hard, big. Fruiting is very good. This variety is popular in Maharastra, U.P., A.P. and Gujarat, up to an elevation of 1000 m.
- ☆ **Dholka:** A popular variety of Gujarat. Trees are vigorous and big. Fruits of big size and good quality.
- ☆ **Nasik:** Fruits are bottle shaped and rough surface. Keeping quality is good.

Propagation

In India and in many other countries, guava is commonly propagated from seed which is quick and cheap method. Seeds retain their viability for one year and germinate within 2 to 3 weeks. Boiling of seeds for five minutes, soaking them in water for about 2 weeks prior to sowing or treating them with strong sulphuric acid for 5 minutes facilitates their germination. The seeds are sown in nurseries or flats at a depth of 0.5 inch. If sown in flats the seedlings are later transferred to containers. When the seedlings are 30 to 45 cm high, they are planted in the field in the beginning of monsoons.

In any way raising plants from seeds is undesirable as it brings variability. Therefore, vegetative propagation of clones is highly desirable. Satisfactory techniques have been worked out for patch budding by the forkert method, side veneer grafting with angular shoots, approach grafting (Inarching), and rooting of cuttings. They may also be grown from suckers which can be induced by severing the roots 30 to 45 cm from the trunk.

Planting

Before planting, it is necessary to have provision for windbreak in South-West direction to protect plants from hot and desiccating wind. It is necessary in North India and coastal area wherever guava is cultured. Most appropriate time for plantation is the onset of monsoon (July/August). However, in irrigated areas it can be preponed to February or in heavy rainfall areas postponed to September to October. For this orchard area should be levelled, cleaned, ploughed and planting place should be suitably marked. Pits of 13 m are made in summer, kept open for a fortnight and filled with a mixture of top soil +50 kg compost +1.25 kg single superphosphate and 150 g Aldrin dust before planting.

Planting distance is governed by the type of rootstock, cultivar, soil fertility, climate, and management practices. However, generally 6x6 m to 8x8 m distance is recommended. In modern horticulture, by using dwarf rootstock and adopting regular pruning, people are opting for closer plantation but the experience has been varied. If planting is done at a closer spacing of (3.6x3.6 m) compared to wider spacing of (7.5×7.5 m), the production was doubled without altering the quality of crop.

Orchard Soil Management

As stated earlier, in the young orchard season wise appropriate intercrop should be taken. However, in a grown up orchard the soil management practices would vary with the type of crop as this fruit gives three crops in a year. In arid conditions, mrig bahar (rainy season flowering and winter crop) is recommended. Therefore, after harvest the orchard is left as such untilled under the cover of natural grasses during summer. In May, orchard is tilled, fertilization is done and the crop is allowed. During rainy season and subsequently until the harvest, competition from weeds should be avoided by clean management as the feeding roots of this crop are very shallow.

Nutritional Management

It has already been mentioned while describing soil that no fertility is excess for guava indicating there by that it responds favourably to manuring. However, in practice, guava does not receive any manure. In Andhra Pradesh, a dose of 100 kg FYM or leaf mould, 3 kg ground nut cake, 3 kg ammonium sulphate per tree are recommended. It is advisable to give a 6 year old are more old tree a dose of 60 kg FYM +2 kg ammonium sulphate, 1.5 kg single super phosphate and 600 g muriate of potash in the beginning of monsoon with half of nitrogen in October.

Irrigation

In guava for high quality fruits, mrig bahar is advocated and encouraged. Therefore, whereas most of the fruit trees would need more irrigation during summer months, these trees can do without it. Therefore, this crop requires two-three post monsoon irrigations during winter at an interval of 15 to 20 days for which rainy season water could be harvested and recycled under arid conditions.

Crop Regulation (Bahar Treatment)

As already stated, depending on the climatic zone where guava orchard is located, a guava tree shall flower twice or thrice in a year. If we leave the tree to nature for it brings in irregularity which is not desirable in commercial orcharding. As already stated, it is desirable to take only one crop in a year which gives better, bulk production and high quality fruits. For this it is necessary to avoid undesirable crop and concentrate on desirable crop. To achieve this, plants are subjected to stress by various treatments to put them to rest during the period of undesirable crop period and forcing them to bear concentrated crop during the period of desirability is referred as 'Bahar treatment'.

The main principle involved in this treatment is that 2 months before flowering period irrigation is withhold so that due to stress leaves would be shed off and plant shall enter quiescence and even some fruits which might have set would also drop off. When part of the leaves had abscissed, soil is cultivated and trees are fed and irrigated for better flowering. In areas where soil is heavy like Maharashtra, orchardists go to the extent of exposing the roots and pruning them. To clarify the matter the method is being detailed below.

We are aware that crop from Ambe bahar (spring flowering) ripens in rainy season has poor quality and even severally infested with fruit fly, which of course is unthinkable in arid conditions. While the produce from Mrig bahar (rainy season flowering) ripens in winter and has high quality with minimum infestation from devastating fruit fly attack and therefore, this crop should be preferred which is only the hope for arid conditions. To eliminate Ambe bahar from January to mid May irrigation is withhold so that flowers, fruits and leaves may abscise and plant goes to rest during summer months. In the last week of May or first week of June, the orchard is deep ploughed, trees are fed 50 kg FYM and irrigated at in interval of 4 to 5 days initially and 10 to 15 days afterwards until the onset of rainfall. Within 20 days of this treatment, trees would flush, bear and give harvest in November to December.

If this treatment is regularly given, the plants are likely to be damaged and their production capacity would dwindle. Therefore, there are some other suggestions by various workers are as follows:

1. Hand deblossoming of spring flowering. It is of course tedious and costly.
2. Spray of growth regulators like NAA at high concentration at full bloom period during Ambe bahar. In this case varied results have been obtained. Application NAA 400 ppm is most effective on 'Banarasi Surkha' cultivar while application of NAA 800 ppm to be affective on 'Allahabad Safeda' cultivar. Application of NAA 80 to 100 ppm and application of NAA 50 and 2,4-D 30 ppm were found to be effective. It seems the effective concentration will require standardization for each variety for each agro climatic zone.
3. Spray of high concentration of urea at 15 to 20 per cent is very effective in eliminating rainy season crop.
4. Spray of Ethephon 1000 to 1500 ppm at full bloom also eliminates the crop and helps in bringing uniform and high quality crop.

Insect Pest and Diseases

Insect Pest

1. **Fruit fly (*Chaetodacus spp.*):** It attacks mostly the rainy season crop. Female lays eggs under the skin of fruits from which maggots emerge within 4 to 5 days, feed through their life of 6 to 20 days on pulp and render them useless. Affected fruits rot and drop.

   Control: To control this, avoid rainy season crop (Ambe bahar) collect affected fruits and dump them in a pit and plough orchard thoroughly during August to September. Spraying of 0.1 per cent Fenthion or 0.08 per cent Malathion regularly provides effective control. However, after spraying fruit harvest should be avoided for 15 days.
2. **Stem borer (*Aristobia testudo*):** Moth lays eggs under the skin from which caterpillars develop and make holes in shoot, and thereby affect the plant health and production.

   Control: To control this pick out the caterpillar with the help of a twisted wire and put a few drops of kerosene oil or Nuvacron or Malathion (0.05 to 0.1 per cent) in each hole and plug it.

3. **Guava scale (*Pulvinaria psidii*):** They are small green insects found on young leaves, shoots and sometimes fruits and damage them by sucking sap.

   Control: To control this, spray of 0.1 per cent Rogor, or Supracid or Meta systox is very effective. Heavily infected shoots should also be pruned out and destroyed.

Besides these, mealy bug which attacks mango and the bark eating caterpillar (*Inderbela quadrinotata*) are also common and same control measures could be adopted. Guava fruits are also attacked by birds like parakeets, crows and bats and animals like monkeys. They have to be scared away.

Diseases

1. **Guava wilt (*Fusarium* spp.):** This is the most serious disease in North India found in alkaline soils. Affected trees show slow browning of leaves which look wilted and the limbs start drying up one by one and eventually the whole tree dies. Causal organism is not fully understood. However, there are reports to indicate association of Fusarium fungus. In Uttar Pradesh, *Fusarium oxysporium* var. psidii and in West Bengal, *Fusarium solani* and also *Rhizoctonia bataticola*, *F. psidii* and *Macrophomia phaseoli* are found associated. It is aggravated by waterlogged conditions.

   Control: Remedy lies in avoiding waterlogged situations, removal and burning of infected trees, and soil drenching with 0.3 per cent Thirum or 0.1 per cent Benlate some scientists feels that application of gypsum, lime and organic manures reduces the extent of damage.

2. **Anthracnose (*Colletotrichum psidii*):** This disease mainly affects fruits, common on rainy season crop. In the beginning rough spots appear on fruit surface which coalesce and become larger, and on storage such fruits rot. This fungus also attacks shoots, flowers and buds and the branches begin to die-back. Parasitic algae (*Cephaleuros mycoidea*), scab (*Pestalotia psidii*) and guava canker caused by bacteria could be other troublesome disease attacking guava.

   Control: It can be controlled by spraying with 3:3:50 Bordeaux mixture or 0.1 to 1.5 per cent Phytolan. For storage only healthy fruits should be chosen. 'Apple colour' cultivar is tolerant to this disease.

Harvesting and Yield

Trees often bear their first fruits 2 years after transplanting, attain full bearing at 8 years and cropping may continue up to 30 years. Under subtropical conditions (North India), guava tree mainly in two seasons: Spring flowering commonly referred as Ambe bahar from which fruits ripen in rains and rainy season flowering (Mrig bahar) from which fruits ripen in winter. However, in tropics, it flowers, three times and the third flowering is in October (Hasth bahar) which ripens in March.

It is desirable to take only one crop in a year which comes in bulk and better for the health of the tree also. In South India, rainy season crop is preferred at the cost of quality for higher price. In the North, winter crop is prefer- red as the fruits escape fruit fly which is more prevalent in rainy season and the crop is of higher quality. For arid conditions only mrig bahar is to be encouraged to take advantage of natural precipitation.

It takes about 5 months for fruits to mature after flowering but prevalent temperature during fruiting may delay or enhance the process of growth. Low temperature would delay the process while high temperature enhances it.

Guava fruit generally ripe 5 month after flowering and harvested when their colour change to yellowish green. The highest quality is obtained in tree ripe fruit but such fruits are highly perishable and theirfore distant market they should be harvesed just at colour break with specific gravity of fruit abouve one. For jelly hard mature fruits (firm ripe) are better. For harvesting hand picking is emplied higher branches can be riched with a ladder. The fruit should be plucked with leafy shoots and collected in bosket. Picking is done 2 to 3 time in a week during harvesting seasoon of 8 to 10 week. The average yield in India is about 90 quental per hectare but yield up to 250 q to 375 q per hectare are also obtained in Allahabad UP.

Chapter 22

Jamun

Botanical name:	*Syzygium cumunii* (Syn. *Eugenia jambolana*)
Family:	Myrtaceae
Chromosome no.:	2n = 40
Origin:	India

Introduction

Jamun is an important but under exploited indigenous minor fruit of India. It is a very common, large, evergreen beautiful tree of the Indian subcontinent. The other common names of jamun are java plum, black plum, jambul and Indian black cherry. The jamun is tall, handsome evergreen tree. It is one of the most popular fruit of India grown throughout the length and width of the country excepting high hills. However, regular orchards of this fruit tree are very rare. It is more popular as an avenue tree and for wind-break. The trees of this species are also found in Thailand, Philippines, Medagasker, West Indies, East and West Africa, Algeria and Israel. In India, it is grown in the Indo-gangetic plains, lower ranges of the Himalayas and Tamil Nadu. It is a very hardy crop and suited very well to arid and semi-arid conditions.

Taxonomy and Morphology

It belongs to the important genus *Syzygium* of the family Myrtaceae to which guava also belongs. There are about 400 to 500 species of which a few provide edible fruits like.

- ☆ ***S. jambos* (Rose apple or Safed jamun).** The tree is ornamental. The fruits are light yellow-white in colour. They are rose scented and calyx persistent. The trees are medium sized and evergreen. The leaves have very small petiole. *S. jambos* is grown in Assam, Bihar, Andhra Pradesh, Tamil Nadu, West Bengal, Coastal areas of Maharashtra and Gujarat. S. seeds are poly embryonic.

- ***S. fruitecosum:*** The trees are suitable for windbreak. Fruits are edible, small in size.
- ***S. javanica* (Water apple):** This species is found in South India and West Bengal.
- ***S. densiflora*:** It is suitable for use as a rootstock for *S. cumunii.* It is resistant to attack of termites.
- ***S. uniflora (*Surinam cherry or Pitanga cherry*):*** It is also a small tree. It bears small sized fruits. The colour of fruit is bright red. The fruits have an aromatic flavour. The tree is found in South India.
- ***S. zeylanica:*** Tree is small. It bears edible fruits. The tree is commonly found in Western Ghats of India. Jamun is an evergreen tree attaining medium to large size. The trunk is thick and greyish white in colour. Leaves opposite. simple, elliptic or obviate, to 5 inch long, pinnately veined with lateral veins close together; flowers in a few flowered panicles; petals white, united into a cap; fruit a berry purplish red, ovoid, to inch long, edible Seed poly embryonic.

Composition and Uses

Quantity of pulp is quite varied depending on variety, of which about 88 per cent is moisture, 0.7 per cent protein, 0.1 per cent fat, 19.7 per cent carbohydrate and 0.4 per cent mineral matter. Fruits are also very rich in iron and calorific value.

Fully ripe fruits are eaten fresh and can be processed into beverages, jelly, jam, squash, wine, vinegar and pickles. Seeds are ground into rough powder and mixed in other concentrate to make animal feed. Wood has timber value and used for railway sleepers being strong and tolerant to insect and fungal attack. The twigs are used as fire wood. The powdered seeds have reputation of being useful in the treatment of diabetes. It is reported to diminish the quantity of sugar in urine very quickly, and in some cases, even permanently.

Climate and Soil

Jamun prefers to grow under tropical and sub- tropical climate. It is also found growing in lower ranges of the Himalayas up to an altitude of 1300 metres. The jamun requires dry weather at the time of flowering and fruit setting. In sub-tropical areas, early rain is considered to be beneficial for ripening of fruit and proper development of its size, colour and taste.

It can be grown on wide range of soils. But for high yield and good plant growth, deep loam and well drained soils are needed. Such soils also retain sufficient soil moisture which is very beneficial for optimum growth and good fruiting. Jamun can grow well under salinity and waterlogged conditions. However, it is not economical to grow jamun on very heavy or light sandy soils.

Important Varieties

There are no standard varieties available in the country and a common type grown in north India is popularly known as '*Rajamun*'. There are no descript varieties available. However, considerable variation exists in fruit size. Another type with large sized fruits, known as 'Paras' in Gujarat, is also popular. The available variation should be subjected to selection for improvement. A selection Narendra Jamun -6, identified with desirable traits at Faizabad, however, is yet to be popularised. Konkan Bahadoli is a seedling selection identified by RFRS, Vengurla.

Propagation

The jamun is propagated both by seed and vegetative methods. Due to existence of polyembryony it comes true to parent through seed and followed in most cases. Though vegetative methods have already attained given success still seed propagation is continued. However, seed propagation is not advisable due to late bearing.

The seeds have no dormancy. Fresh seeds can be sown. Germination takes place in about 10 to 15 days. Seedlings are ready for transplanting for the use as rootstock in the following spring (February to March) or monsoon *i.e.* August to September.

Propagation of jamun by budding is economic and convenient. Budding is done on one year old seedling stocks, having 10 to 14 mm thickness. The best time for budding is during July to August in low rainfall areas. In the areas where rains start early and are heavy, budding are attempted early in May to June. Among methods of budding, shield, patch and forkert have proved very successful. How- ever, results of experiment conducted at Fruit Research Station, Basti, Uttar Pradesh showed 70 per cent success by patch budding when done in the month of March. The possibility of better success has been reported in forkert method comparing to shield or T-budding.

Jamun can also be propagated by inarching but it is not adopted commercially. In this method, one year old seedling raised in pots, are inarched with the mother jamun trees with the help of wooden stands during June to July.

About 60 per cent air-layers are obtained with 500 ppm IBA in lanolin paste is used, provided air-layering is done in spring and not in the rainy season.

Better rooting through cuttings is obtained in Jamun under intermittent mist. Semi-hardwood cuttings of both *S. jambos, S. javanica*, 20 to 25 cm long, taken from the spring flush and planted in July treated with 2000 ppm IBA give higher percentage.

Planting

Prior to planting, field is properly cleared and ploughed. The pits (1 x 1 x 1

m) are dug at the distance of 10 m both ways. Usually, work of digging of pits is completed before the onset of monsoon. The pits are filled with the mixture of 75 per cent top soil and 25 per cent well rotten farmyard manure or compost.

Jamun is an evergreen tree and can be planted both in spring *i.e.* February-March and the monsoon season *i.e.* July-August. The transplants are taken out along with whole root system with a ball of soil, and are planted in the centre of the pit.

Orchard Soil Management

During rainy season green manuring could be practiced, otherwise for rest of the period of the year soil should be left undisturbed to prevent soil loss through erosion.

Training and Pruning

Training of young plants is essential to provide better frame work. Keep the main stem or trunk clean up to a height of 60-90cm from the ground level by removing the basal branches and sprouts where sprout emerging from the rootstock portion should be periodically removed. During latter part of plant growth jamun plants do not require any pruning except removing diseases, weak, dry crisscross branches/twigs.

Nutritional Management

About 20 kg of well rotten farmyard manure or compost should be applied per plant per year. For bearing trees this dose is increased to 50 to 80 kg depending upon the size and the age of tree. This much quantity of organic matter suffices the requirement for growth, flowering and fruiting. In those bearing trees which are growing luxuriantly, the application of manure and fertilizer may be avoided.

Irrigation

The well developed deep tap root system of jamun is able to extracted water from deeper layer of soil that allows the plant to thrive well under rain fed condition. Usually, 8 to 10 light irrigations are required during the early age of the plants. Bearing plants are hardy in nature. However, sufficient soil moisture is required during fruit ripening stage. Plants can be protected through irrigation during severe winters from frost damage. Bearing trees should not be given irrigation in September to October for better fruit bud formation. Similarly withholding irrigation in February to March enhances flowering and fruiting depending upon climatic and edaphic conditions.

Flowering and Fruiting

Flowers are borne in the axils of leaves on branchlet. In North Indian conditions, flowering starts in the first week of March and continues up to the end of April. The maximum anthesis (18.71 to 43.08 per cent) and dehiscence are between 10.00 and 12.00 a.m. The pollen fertility is higher in the beginning of the season. The maximum receptivity of stigma is one day after anthesis.

The jamun is a cross pollinated crop. The pollination is done by honeybees, houseflies and wind, maximum fruit set (32.6 to 36.0 per cent) can be obtained by hand pollination when it is done one day after anthesis. Thereafter, a sharp decline is observed in fruit set.

There is heavy drop of flower and fruit within 3 to 4 weeks. It is as high as 50 to 60 per cent. A large number of fruits drop off at very young stage during 5 to 8 weeks after blooming. Later, natural fruit drop declines. The flower and fruit drop can be reduced with two sprays of CA, 60 ppm, one at full bloom and other 15 days after initial setting of fruit.

The pattern of growth and development of jamun fruit has been divided into three phases *i.e.* the first phase from 15 to 52 days after fruit set having slow growth of fruit, the second phase from 52 to 58 days after fruit set having fast growth and the third and last phase from 58 to 60 days after fruit set having slow growth, and very little addition in fruit weight. Studies conducted at Horticultural Experiments and Training Centre, Chaubatia, U.P. on pollination, fruit set and fruit development in jamun *cv* 'Frienda' recorded maximum fruit set under open pollination which is affected by wind, insects and gravity. Maximum fruit drop was recorded to be 84.63 and 81.81 per cent in 2 consecutive years. The fruits took about 3.5 months to mature after full bloom.

Insect Pest and Diseases

Insect Pests

1. **White fly (*Dialeurodes eugenia*):** Affected fruits gets wormy appearance on the surface.

   Control

   a. Maintain sanitary measures around the tree in the orchard.

   b. Pluck all affected fruits and destroy them.

   c. Dig up the soil around the tree trunk so that the maggots in the affected fruits and the pupae hibernating in the soil are destroyed.

2. **Leaf eating caterpillar (*Carea subtilis*):** Caterpillars attack the leaves and tree becomes defoliated.

   Control: Spray Rogor 30 EC 0.1 per cent or Malathion 0.1 per cent.

Besides the above insects, the Jamun crop is seriously by pests like squirrels and birds like parrots and crows, these have to be frightened away by beating the drum or flinging stones.

Diseases

1. **Anthracnose (*Glomerella cingulata*):** The fungus incites leaf spot and fruit rot. Affected leaves show small scattered spots light brown or reddish brown in colour. Affected fruit show small, water soaked, circular and depressed lesions. Fruits rot and shrivel.

Control: Spray with Dithane Z-78 at 0.2 per cent or Bordeaux mixture at 4:4:50 concentrations shall check the disease.

Physiological Disorder

Flower and fruit drop- in jamun the flower and fruit drop start just after opening of flower and continue up to maturity. About 65 per cent flower and fruit drop in the first five weeks and since then a maximum of 19-21 per cent flowers and fruits drop off up to maturity. Only 12-15 per cent flowers reach maturity.

Control: It may be reduced by two sprays of 60 ppm GA_3 one at full bloom and the other 15 days after initial setting of fruit.

Harvesting and Yield

On ripening fruits change their colour into deep red or violet. Usually, fully ripe fruits are plucked by shaking the branches by climbing on the tree. Fruits are not allowed to fall on the ground due to fear of getting damaged or bruised and smeared with dust. Since all the fruits are not ripe at a time, 3 to 4 harvestings are needed to complete the process.

The fully grown seedling Jamun tree produces on an average 80 to 100 kg of fruits whereas it is about 60 to 70 kg from vegetatively propagated tree.

Chapter 23

Kair (Karira)

Botanical name:	*Capparis decidua*
Family:	Capparidaceae
Chromosome no.:	2n = 14
Origin:	India

Introduction

Kair is a fruit plant grown in wild state in arid and semi-arid region of the country. It is generally used for pickles and sometime as vegetables. If proper selection is made from available variability it makes an excellent crop for extreme arid of Rajasthan and Gujarat.

Taxonomy and Morphology

Kair is belong to family Capparidaceae and genius capparis which consist of about 270 species of which 40 species are wild in India. The plant is densely branching shrub with green, thorny stem and scanty, small caduceus leaves found only on young shoots. Red scarlet flowers appear in March to April. Berries of baring size 1-1.5 cm diameter, globose, green when immature turning to red on ripening and have sweet slimy pulp.

Composition and Uses

Fruit pulp is about 55 per cent of total weight which contain 56 per cent moisture, 8.6 per cent protein (crude) 5.02 per cent true protein, total sugar 1.8 per cent reducing sugar 1.1 per cent Phosphorus 0.057 per cent Potassium, 1.026 per cent Calcium, 0.055 per cent Magnesium and ascorbic acid 7.18 mg per 100gm. Kair fruits are mainly used in pickling or as vegetables when immature.

Ripe fruits are sweet and acrid and enjoy by children in natives. Fruits are useful in cardiac trouble and biliousness, tender leaves and shoots are used for plastering boils. Bark is useful in cough, asthma and inflammation. Root bark is used against intermittent fever and rheumatism. Immature fruits are even dried and used as vegetable in off season.

Climate and Soil

It can be grown in dry climate of tropics and subtropics and therefore wide spread in deccan platu Rajasthan and Punjab. It is highly suitable for arid and Semi-arid region.

Light sandy soil where most of the crops would fail to grow. Though it is a very hardy tree but plants do not survive under waterlogged conditions. It can grow even in pockets of soil between crevices of barren rock. Trees are grown on degraded rocky area including salt-affected soils. However, for its better growth and productivity, well-drained deep loam soil is ideal. It prefers tropical and subtropical climate and can withstand drought admirably.

Important Varieties

Since it exists only in semi wild fall no varieties are known however great variation of fruit size and pulp content exist.

Propagation

It can be propagated by seed or through suckering which in common in this crop.

Planting

No regular planting of kair is known however looking to it size and bushy nature in the event of regular plantation in a field or on the boundary as a compact hedge or for sand dune stabilization plantation can be done at (3×3 m).

Nutritional Management

A fully grown, bearing plant should be top dressed with fertilizers @ 100 g N, 160 g P_2O_5 and 260 g K_2O. These may be applied in two splits, in June-July and again in January- February, when there is sufficient moisture in the soil.

Irrigation

Young plants need to be cared for watering time to time in the first year of planting and when plants establish well, usually watering is not done. However, sufficient moisture helps in good yields for which runoff collection should be practiced in rainy season by developing micro catchments around the trees.

Flowering and Fruiting

Normally kair flowering take place twice in a year *i.e.* in March to April and

July to August after the plant had attained an age of 5 year. From March to April flowering the green immature accrued fruits are harvested during May to June for pickling, dehydration and vegetable purpose. The fruits from second flowering are available in October which is rarely used. A five year old bush gives about 5 kg fruits.

Insect Pest and Disease

There is no insect pest and disease which are damaged economically.

Harvesting and Yield

Fruits are harvested during May to June for pickling, dehydration and vegetable purpose. A five year old bush gives about 5 kg fruits.

Chapter 24

Karonda

Botanical name:	*Carissa carandas*
Family:	Apocynaceae
Chromosome no.:	2n= 22
Origin:	India

Introduction

Karonda (Christ's Thorn) is an evergreen spiny shrub or small tree. It is widely grown in India, South Africa and Malaya. In India, it is said to be wild in Bihar, West Bengal and South India. It is an excellent hedge plant quite frequent in Punjab and Gujarat. It flowers from February to April and makes use of natural moisture availability period and therefore, highly suitable for arid region.

Taxonomy and Morphology

A genus of densely branched, spinous, erect shrubs belonging to Apocynaceae family. There are some twenty species of which 5 are of Indian origin including *C. Carandas*. This is a dichotomously branched spinous bush. Leaves opposite, small coriaceous; flowers are white fragrant and born in clusters of 2 to 3. Corolla tube cylindrical, naked. Anthers included, free from the stigma. Ovary 2-celled, fruit a berry ellipsoid or globose which is grape-green when young, white and pink when approaching maturity and nearly black when ripe.The fruits ripe in July to August.

Some other cultivated species of Carissa are as follows:

- ☆ ***Carissa grandiflora (Natal plum)*:** The fruits are dark red with papery skin and a few small seeds. Fruits are available throughout the year.

- ***Carissa bispinosa*:** The flowers are small and fruits are borne in clusters. They resemble to natal plum. The seeds are lanceolate.
- ***Carissa edulis***: Flowers are white, pink and scented. Flowers are borne in axillary clusters of 10 to 25. Fruits are oval in shape, they are red in colour but turn black at maturity.
- It is a straggling shrub.
- ***Carissa ovata:*** Fruits are small and are used for jam making. The origin place is said to be Australia.
- ***Carissa spindrun***: This species bears small fruits. It is native of India. This is the hardiest species of Carissa and is known to thrive in poor and rocky soil.

Composition and Uses

Karonda fruit is a rich source of iron and contains a fair amount of Vitamin C. Mature fruit contains high amount of pectin and minerals. Fruits are eaten fresh by children but more useful in pickling, chutney and for culinary purposes. It also makes excellent jelly. Fruit is considered to be anti scorbutic. The fruit juice is irritant and capable of producing ulcers. The ripe fruit is pleasant acid, and goes well with food. The plant makes excellent strong hedge almost impassable.

Climate and Soil

Karonda is very hardy in habit. It can grow in tropical and sub-tropical climatic conditions. However, heavy rainfall and waterlogged conditions are harmful. Temperature climate with high frost and snowfall areas are not suitable for this fruit. Plants are sensitive for low temperature and cold injury.

Karonda can be grown on a wide range of soil including poor soils of arid. However, on good soil, its growth is better and fruit yield is higher. It can be grown in wide ranges of soil pH ranging from 5.0to 8.0.

Important Varieties

On the basis of fruit colour, the varieties of karonda can be classified as:

i. Green fruited,

ii. Pink fruited,

iii. White fruited.

The differences are not much in shape and size of fruits from all the three varieties. However, there is tremendous scope for improvement using selection force and vegetative methods of propagation.

- **Pant Manohar:** This variety is developed from GB PUA and T Pantnagar (Uttarakhand) in 2007. The plants of this varieties are medium– sized dense bushes, fruits are dark pink blush on white background.

- ☆ **Pant Sudarshan:** This variety is developed from GB PUA and T Pantnagar (Uttarakhand) in 2007. The plants of this varieties are medium– sized dense bushes. Fruits are pink blush on white background. On ripening fruits become dark brown.
- ☆ **Pant Suvarna:** This variety is developed from GB PUA and T Pantnagar (Uttarakhand) in 2007. Plants are upright growing and sparse. Fruits are dark brown blush on green background.
- ☆ **Konkan Bold:** This variety is developed from KonkanKrishi VidyaPeeth, Dapoli (Maharashtra) in 2004. The plants are medium in size and vigourous. It flower in the month of Feb.-March and fruit ripe in the month of May-June under Coorg conditions. The colour of fruits is dark purple.

Propagation

Karonda is commonly propagated by seeds and vegetative methods like stem cuttings, inarching and air- layering. The fresh seeds are sown in the nursery in the month of August or September. Under good management, seedlings become ready in a year's time for transplanting.

Karonda can be propagated by hardwood cuttings successfully. Though cuttings are difficult to root type, they may also be planted under polythene tent which gives higher rooting because under such structure the humidity is high. Where mist facilities are available, soft and semi-hardwood cuttings can be planted for successful rooting and rooted cuttings are transferred to pots containing good growing media consisting of two- third of garden soil and one-third of well rotten compost. After 6 to 8 months, plants attain transplantable height. Soft wood, grafting is also successful in karonda and is very useful technique for in-situ propagation in arid.

Air-layering is quite successful in karonda. It is done in the beginning of monsoon. Rooted layers are obtained after 3.5 months of layering. Propagation of karonda has also been tried by inarching which is successful. However, compared to other methods inarching is not very popular.

Planting

The field is cleaned and leveled for layout. The pits of 30 to 45 cm^3 are dug and filled with organic manure and soil in the ratio of 1:1. The distance of planting in hedge is recommended at 1 to 1.5 m and regular planting at 1.5 to 2.0 m both ways. The best time of planting is the beginning of monsoon.

Orchard Soil Management

In initial years competition from weeds should be kept to minimum. During rainy season green manuring crops are suggested and during rest of the period no tillage is recommended excepting the period of manuring, which is affected prior to onset of monsoon.

Nutritional Management

Balanced nutrition is considered to be the most important which determines productivity and quality. Karonda is a hardy plant. Regular plantations are very rare because it is preferred to be grown as protective hedge which is generally uncared. Being a minor fruit crop scientific experiments were not conducted on manurial requirements. However, reports from progressive cultivators reveal that karonda plants respond well to organic manuring when applied at the rate of 15 to 20 kg and 400 g of mixture of NPK per bearing plant per year preferably in the beginning of monsoon period or the plant will slowly get exhausted due to continuous bearing and finally dieback and die.

Irrigation

Karonda is comparatively a slow growing plant, once established will not need much water. Moreover, karonda behaves as a xerophytic plant therefore, water requirement is very low. However, efforts should be made to collect runoff during rainy season in root zone through land shaping.

Flowering and Fruiting

Karonda starts bearing flowers and fruits from third year of its planting. The plant starts bearing flower in the month of February and ripe fruits are available in the month of August. However, unripe fruits are available from May onward. Under arid condition, leaving it to nature, its flowering will coincide with monsoon and fruits will be available in post- monsoon period.

Insect Pest and Diseases

Insect Pest

1. **Leaf eating caterpillars**: Caterpillars cause much damage, mainly by eating leaves. This affects the growth of the plants. Caterpillars may be controlled by the use of pesticides, biological control and cultural practices. The chemicals, monocrotophos (2ml/L may be used for control of leaf eating caterpillars.
2. **Fruit fly (*Bactrocera dorsalis, B. caryeaea*)**: Fruit fly infests the ripened fruits. Its infestation is more in southern states. The female fruit fly lays eggs on the mature fruits with the help of its pointed ovipositor. After hatching the maggots feed on pulp of these fruits and the infested fruits starts rotting and fall down. As a result brown patch appears around the place of oviposition. The maggots come out of the affected fruit and pupate in the soil. Pre-harvest IPM combined with sanitation (Collection and destruction of fallen/infested fruits) + Placing Methyl eugenol trap @ 4-6/acre + In severe infestation spraying of bait spray (Decamethrin (Decis) 2ml+ 100g of jaggery in 1 litre of water) is recommended.

Diseases

1. **Anthracnose (*Colletotrichum inamdarii*):** This disease is wide spread in U.P. and can be seen in the form of black- greyish spots on the leaves.

   Control: Spray of 0.1 per cent copper sulphate can control this disease.

2. **Bacterial canker (*Xanthomonas carissa*):** Symptoms first appear as spots on lower surface of leaf. These spots are small, round and water soaked turning dark brown. A yellowish green halo may surround the spots which may cause necrosis. Upper surface of the leaves look raised behind the spots.

   Control: The disease can be checked by removing diseased leaves followed by a spray of (phytomycin 200 ppm).

Harvesting and Yield

Karonda plant starts yielding after 3rd year. In Western Ghats flower starts in December to March and fruit mature in the month of April to June. Both unripe and ripe fruits are harvested. Complete harvesting for all fruits may not be done. It is done in two to three times. Colour change is a good indicator of maturity of fruit. About 4 to 5 kg fruits per bush are obtained.

Chapter 25

Lasoda

Botanical name: *Cordia myxa*

Family: Boraginaceae

Chromosome no.: 2n= 14

Origin: India

Introduction

Lasoda is also known as Gonda or Lahsua. It is very common fruit tree found grown all over India accepting high hills. It has great capacity to tolerate drought once established and therefore quite wide spread in semi-arid and arid region of North India.

Taxonomy and Morphology

It is an important member of family boragenaceae and genous Cordia to which large no. of species belong of which cordia gharaf is another popularly grown species the plant is small or medium size tree which a short usually crooked trunk. Leaves large simple, alternate 3-5 nerved at the base not tomentose beneath. Flowers white axulary in bunches. The fruit is drupaceous 0.5-1 inch long yellowish brown, pink or nearly black when ripe with a viscid, sweetish almost transparent pulp surrounding the central stony part. Fruit is edible sweetish and mucilaginous.

Composition and Uses

Fruit has mucilaginous pulp which constitutes about 70 per cent of the fruit. It is rich in sugar but acid less. Unripe fruits are used in pickled and mixed pickled with mango is excellent in taste. Ripe fruits are eaten raw or made into spirituous liquor.

Climate and Soil

It thrives well under sub-tropical and tropical condition up to the elevation of 5000 feet. It needs warmer climate and sensitivity to frost. It can stand drought to a very greater extent. Not very exacting in soil requirement with great capacity to grow in stress due to water or salinity. Under tropics flowering can be regulated through moisture availability while in sub-tropics temperatures is the determining factor.

Important Varieties

Since it has been propagated through seeds, there are no well defined cultivars. However, great variation exists in the size of fruits and their pulp content. Accordingly large fruited types in Gujarat are recognized as 'Paras' Gonda which in fact is general term for any fruit variety with big size fruit. There is a need for survey, evaluation, selection and commercialization through vegetative method of propagation.

Propagation

It is generally propagated by seed. Seeds from freshly harvested ripe fruits are extracted and sown in the months of April to May. Plants are ready for planting in July to August. Budding can be successfully done on seedlings during July to September for which one year old seedlings are suitable. *Cordia gharaf* can also be used as rootstock.

Planting

Under arid conditions appropriate time for planting is the beginning of rainy season for which pits of 60×60×60 cm size are made at 6×6 m apart during summer. These pits are refilled with soil and FYM (1:1) just before planting period. It would be advisable to create small catchment around each plant to harvest runoff during rainy season.

Nutritional Management

No manuring is practiced. However, for good and regular production fertilization shall be beneficial. Each tree can be given 25 kg FYM and 500 gm of nitrogen in the beginning of monsoons. However, exact requirements will require investigations.

Irrigation

In the event of failure of monsoon in the year of planting, irrigation is required for young saplings for establishment. Once this is achieved, irrigation is not required for this plant. However, in tropical arid, the flowering can be monitored by the application of water and thus fruiting can be regulated.

Flowering and Fruiting

A graft comes into bearing in fourth year and flowering takes place in March to April. The fruits ripe in May to July. In tropical conditions the crop is comparatively early and in sub-tropics it is late depending on the period of flowering.

Insect Pest and Diseases

Since regular orcharding is rare no serious efforts have been made to record the insect pest and disease problem. However, it is an excellent host for lac insect which may cause economic loss of crop. Secondly leaf gall midge causes serious damage to plant leaves which can be seen in the form of pustules on under surface finally leading to severe defoliation. This can be controlled by a spray of 0.05 per cent Monocrotophos or any other systemic insecticide.

Great variability in respect to fruit size and pickling quality exists in nature which requires to be exploited in making this as a viable fruit crop of arid region in India.

Harvesting and Yield

Cordia fruits are mostly used for pickling or culinary purposes and therefore, harvested green mature. Whole bunch of fruits is harvested and marketed. In case ripe fruits are to be harvested for processing (for liquor), this can be done on the change of color to pinkness. Each grown up tree can yield about 100 to 125 kg per season.

Chapter 26

Mulberry

Botanical name:	*Morus* spp.
Family:	Moraceae
Chromosome no.:	2n= 308
Origin:	Northern china/Japan

Introduction

Mulberry is one of the most common fruit tree found all over India from temperate to tropical. It is mainly cultivated for rearing silk worm but its importance as fruit crops should also not to be undermine it is one of the rare delicacies. However it has not been able to attain commercial significance as a fruit crop due to its highly perishable nature. However owing to its ability to grow under wide variety of agro-climate. It can get into arid and Semi-arid region of the country. Mulberry introduce in India from Persia It is widely cultivated in South India.

Taxonomy and Morphology

It is an important genus of family Moraceae. There are 10-12 species of which *Morus abla* is cultivated for silk worm rearing and *Morus rubra* and *Morus nigra* for their fruits. It is a deciduous, monoecious tree is attaining a height of 10m or above. Leaves are simple, alternate often lobed usually crenate to dentate glabrous to scabrous, flowers unisexual in druping catkins, fruit a juice syncarp and aggregate of drupelets with pink red or black color depending on the species.

Composition and Uses

Mulberry fresh fruit consist of 88 per cent water and only have 60 calories per cup (140 g) and 10 per cent carbohydrates. The fruits are consumed fresh or processed into Jam, beverages and wines. Dry fruit ground to powder make excellent poultry food. Fruit juice has therapeutic value against sore throat, fever, dyspepsia and melancholia. The bark of tree is used in paper industry and leaves are used as cattle feed. Roots have the medicinal value being anthelmintic. Fruit is very juicy and rich in sugar about 10 per cent and iron.

Climate and Soil

Mulberry is basically a temperate region tree but it grows as well in subtropical and tropical climatic condition. It can tolerate drought and cold equally well.

The mulberry tree is very hardy it can be grown on poor soil even in rocky soil. However clay loam or sandy loam soil is better.

Important Varieties

There are no descript varieties is cultivation in India however some varieties are introduce from other country like EC-120294 (Ramania), Back English, Wellington mulberry, Tout Amer, Variegala globosa from (France). The important (Indian) varieties are Kosen (Japanese Variety) Kanva-2, S-30 and S-54 (More popular in Karnataka, Tamil nadu and Andhra Pradesh), S-1 and S-799 (West Bengal).

Propagation

Mulberry is propagated by vegetative method because the plants raised by vegetative methods are true to type and grow uniformly. Propagation by stem cutting is easy and then by budding and inarching. Stem cutting are prepared from mature shoot 20-25 cm long having 3-5 bud. The time of planting of cutting is in the month of December-January. This period also coincide with pruning operation. The pruned shoots are used for propagation. Cutting start sprouting in March and are ready in transplanting during rain. The plant raised by cutting may also be used as root stock and budding can be done with superior type.

Planting

On well cultivated leveled weed free soil of 60 cm are dug during October to November at a distance of (60×60×60) are dug during October to November at the distance of (6×6 m) the pith should be filled with the mixture of FYM and Orchard soil with the ratio of 1:3.

Training and Pruning

Young plant requires some support so that they may grow erect. Pruning is an essential operation under which all shoots on the tree at a height of (1-1.5 m) from the ground level are pruned in December or January when they have shed the leaves.

Nutritional Management

An grown up tree 30 kg FYM, 50 gm N, 40 gm P and 40 gm K should be applied in the month of December in a year.

Irrigation

Though mulberry is drought resistance it responds to irrigation which result in better yield under arid condition it would be desirable to collect runoff water in root zone.

Flowering and Fruiting

Mulberry flowers in February and the crop mature in May to June in Northern India. The trees are monoecious and fruits are aggregate of drupelets.

Insect Pest and Diseases

Since the crop has no popularize as fruit crops. There are not many records of disease and pest. However powdery mildew (*Phyllactinia corylea*) is common in all the mulberry growing areas. It can be controlled by Bordeaux mixture. Mosaic has also been reported in some areas by slight curling and puckering of leaf. It is transmitted by *Rhopalosiphum moidis* and *Myzus persicae*. It can be controlled by use of resistance variety and application of systemic insecticide.

Harvesting and Yield

Mature fruits are harvested on the judgment of color change. The harvesting is over in 3-4 picking. A fully grown up tree can yield 10-15 kg of fruits.

Chapter 27

Phalsa

Botanical name:	*Grewia subinaequalis*
Family:	Tiliaceae
Chromosome no.:	2n= 18
Origin:	India

Introduction

Phalsa is a shrub or small tree of Indian origin. It has been used since Vedic period. The ripe fruits are consumed fresh or processed into refreshing fruit and soft drinks during summer. Phalsa fruit has a short shelf life suitable only for local marketing.It is cultivated on a commercial scale mainly in the northern and the western states of India. Phalsa is known by different name in different Indian languages such as Phalsa, *dhamin*, *parusha*, and *shukri* in Hindi, *dhaman* in Punjabi, *man-bijal* in Assamese, *phalsa* and *shukri* in Begali, *mirgi chara* and *pharasakoli* in Oriya, *phalsa* in Gujrati, *phalsi* in Maharashtra, *jana, nallajana, phutiki* in Telagu, *palisa, tadachi* in Tamil, *buttiyudippe* and *tadasala* in Kannada.

Phalsa is found wildly growing in UP, Rajasthan, Punjab, Haryana, MP, West Bengal and many parts of south India. The cultivation of phalsa is limited to very small scale in Punjab, Haryana, Rajasthan, and Gujarat and UP. Most of the commercial plantations are adjoining to big cities. Apart from India it is cultivated in Pakistan, Nepal, Bangladesh, Laos, Sri Lanka, Thailand, Philippines, Vietnam and experimental basis in some provinces of United States of America.

Taxonomy and Morphology

Phalsa belongs to *Tiliaceae* family. The botanical name of phalsa is *Grawia subinaequalis* DC which was earlier known as *Grewia asiatica* L The closet relative of phalsa is *Grawia elastica* var. *vastita.*which is generally found in small hillocks and valleys. Apart from these two species, there are around 140 species in genus *Grewia.* The important ones are *Grewia glabra, Grewia micrococas, Grewia optiva, Grewia tilifolia,* and *Grewia belosa.*

The phalsa plant is a medium, drooping shrub, which may attain a height of 4 m if unpruned. The stem is gray in colour with rough bark. It bears several long, slender, drooping branches. The young branches are covered with dense hairs. The leaves are alternate, simple, and broadly cordate to ovate shaped with pointed tip. The leaves are 20 cm in length and 15 cm in wide with coarsely toothed with 1-2 cm petiole. The inflorescence is 3 to 5 flowered axillary cymes clustered in groups of 2 to 8 which are 16 to 25 cm long. The receptacle is 3 to 4 mm with hairy upper half. The flowers are small 1 to 2 cm diameter and bright orange-yellow in colour. There are five 5 oblong, 1-5 cm long glabrous sepals. The petals are also five in numbers but 5 to 7 cm long and orange -yellow in colour.

The fruits are very small, purple to crimson red in color when ripe. The fruits have 2 to 3 cm long peduncle and are produced in clusters. Fruit is botanically a drupe. Fruits are 1.0 to 2.0 cm in diameter, 1.0 to 1.5 cm in length with average fruit weight of 1.0 to 2.0g. Fruits ripe 45 to 55 days after flowering, all fruits on a cluster do not ripe at a time. At ripening the fruits turn from light green to cherry red or purplish red and finally in dark purple. The ripe fruit is covered with a very thin, whitish blush, and becomes soft and tender. The flesh is light greenish-white becomes colored purplish-red near the skin. The fruits are sweet acid in taste with mild pleasant flavour. The fruit usually contain one seed, which are hemispherical and 5 – 7 mm in size.

Composition and Uses

Fruits are rich source of vitamin A and C. They are fair source of phosphorus and iron. Fruit contain 50-60 per cent Juice, 6.8 to 25.8 per cent Carbohydrate, 10-11 per cent Sugar and 12 per cent TSS with 2-2.5 per cent acidity. The phalsa fruits are rich in flavonoids, carotenoids and anthocyanins. Flavonoids have known antioxidant activities while anthocyanins help to reduce the risk of heart disease by inhibiting cholesterol formation. The phalsa fruits are rich in potassium, which plays important role in energy metabolism and normalizing blood pressure. Fruits are used as table purpose by children otherwise basically it makes an excellent juice and squash.

Climate and Soil

It grow under sub-tropical climate but it does best in region with distinct winter and summer, where in winter it sheds leaves and fruits in summer. In tropics continuous and irregular crops of poor quality is a common site but fruiting can be monitored through moisture regulation. A grownup plant can tolerate freezing temperature for a few days. The plants can tolerate temperature as high as 44 °C. Clear weather during fruiting enhances the quality of fruits.

It can be grown in wide variety of soil and considered one of the most hardy fruits and drought resistance. For best result well drained loamy soil are preferred. Under water stagnation plants become chlorotic. They are moderately sensitive to salinity and in calcareous soil iron chlorosis is a common problem.

Important Varieties

There are no distinct varieties available in phalsa. Peoples have given some local names "Local" and "Sarbati" no much variability is available. However bush type (dwarf) and upright growing habit (tall) clones are available. The dwarf variety has higher total sugar and non –reducing sugars while tall variety has more reducing sugar.

Propagation

Phalsa is mostly grown from seeds. However propagation from cutting with the use of plant growth regulator (IBA) is possible. The greatest rooting in phalsa cutting by ringing the shoots 15 days before planting and using mid portion of the shoots and treating with 2500ppm IBA.

Fleshly extracted seed are used for raising seedlings as they slowly loose viability with age which under ordinary condition is around 90-100 days. It takes 15 days for seeds to germinate and seedlings are ready for transplanting within three to four month. Seed can be kept fresh for longer period in polythene bag. The soft wood grafting techniques has been found most successful in phalsa.

Planting

Planting of phalsa can be done in July –August or February. Those areas where phalsa tree shed their leaves in winter and plant remains in dormancy; February is best time of planting. In those area where winter is very short or there is no winter, phalsa plants do not shed leaves, rainy season *i.e.* July-August is the best time of planting.

The planting distance is usually recommended 2.5-3.0 m both way but closer planting can also be affected as fruiting season is very short and the plant can be subjected to heavy pruning. For planting of phalsa the pit size is (75x75x75 cm).

Training and Pruning

Pruning is an important cultural practice which is an annual cultural function in orchard management. Practices of pruning are quite varied in various part of the country. Phalsa plant is allow to develop as bush therefore no initial training is required. In general pruning is in the middle of December to a gave maximum production. Phalsa bears fruit on current season's growth and for good yield, there is a need for regular annual pruning to cut the old growth and to enhance the new growth. Annual pruning to a height of about 0.75 to 1.25 m encourages new shoots and higher yield of marketable fruit than does more drastic trimming.

Nutritional Management

Being a hardy crop fertilization is hardly practices but since the bearing is on new growth it would definitely respond to good fertilization. One kg nitrogen per plant to be sufficient for good crop. For one hectare crop of phalsa require 100 kg nitrogen, 40 kg phosphorus and 25 kg potassium per plant is required. Phalsa plant is very sensitive to Iron deficiency. Therefore spray of 90.4 per cent) Iron sulphate would be beneficial. Spray can be made at pre bloom and post bloom stage.

Irrigation

Phalsa is regarded as draught tolerant crop. However it has been observed at irrigation is necessary for quality and big harvest. Irrigation at 2-3 week interval during summer or desirable.

Flowering and Fruiting

Phalsa plant start bearing in a year or two but comes to maximum capacity in third year and their after. Flowering start in spring season which continue for about a month with the extension of new growth. Fruits begin to ripe in hot summer starting from March in South to May end in North India. The fruits are small in size and they ripen over an extended period 28-31 days. To reduce the duration of ripening and to improve the size of fruit a no. of plant growth substances can be use like GA3- 60ppm and Urea 1 per cent during flowering period reduce the duration of harvest while GA3 60 ppm + 2,4,5,T ppm is effective in increasing the fruit size and yield.

Insect Pest and Diseases

Insect Pest

1. **Mealy bug (*Drasicha mangiferi*):** It is severely affecting the fruit setting it can be controlled by application of Diazinon or Monocrotophas 0.04 per cent).
2. **Bark eating caterpillar (*Inderbela tetranicx*):** It is poly phagus insect which damage the plant by making the tunnel in the main branches or trunk. It can be controlled by injecting Kerosene oil or Petrol in the holes and plugging the mouth with mud.
3. **Leaf eating caterpillar (*Euproctis fraterna*):** This caterpillar feed gregariously on leaf lamina skeletonizing the same completely. In case of sever infestation entirely may be defoliated. It can be controlled by application of Carborile or any systemic insecticide @ 2-5 ml per liter of water.

Diseases

1. **Leaf spot disease (*Cercospora greviae*):** On affected leaf tinny brown lesions appear on both side of the leaves and cause premature leaf fall,

particularly during rainy season. It can be control by application of (Blitox 0.2 per cent or Dithen-Z 78.3 per cent).

2. **Rust (*Dasturella grewiae*):** Light brown spot develop on the lower side of leaves as a result of the infection. This results in defoliation of leaves. Alternate sprays of Dithen M –45 (0.3 per cent) and Sulfex (0.2 per cent) at 15 days interval effectively control this disease.

Harvesting and Yield

Fruits are harvested when ripe which can be judged by development of dark red color. Such fruits are individually picked and collected in small bamboo bosket cuchioned with soft grass. Since fruit ripe over a period of one month picking has to done every alternate or third day. The average from a mature bush is estimated to be 5-10 kg.

Chapter 28

Pomegranate

Botanical name:	*Punica granatum* L.
Family:	Punicaceae
Chromosome no.:	2n=16, 18.
Origin:	Iran (Persia)

Introduction

Pomegranate is a favourite table fruit of tropical and subtropical region of the world. It is extensively cultivated in Mediterranean countries since age's mainly in Spain, Morocco, Egypt, Afghanistan, Baluchistan and to some extent in Myanmar China, Japan and California (U.S.A.). In India trees are found growing from Kanyakumari to Kashmir but commercial plantation exists in Maharashtra and Gujarat and to a limited extent in U.P., A.P., Karnataka and Tamil Nadu. Correct estimate is not available but in last one dictate in Maharashtra area has come to about 25,000 ha and it is picking up in Gujarat also.

This fruit crop can tolerate salinity and saline irrigation water and does well even in shallow soils. It can also tolerate drought to some extent and fruiting can be tailored to the periods of soil moisture availability. Therefore, it makes an excellent choice under arid semi-arid conditions and has great future in India.

Taxonomy and Morphology

Pomegranate belongs to a small family Punicaceae. This genus *Punica* has 3 to 4 more species but *granatum* is only important economically. This has again been sub divided into chlorocarpa and porphyrocarpa sub-species. Plant is hardy and bushy taking a height of 2 to 4 m and is deciduous in colour part of its range. Leaves opposite, shorts petiole, glabrous, shinning, dark green, oblong, 4 to 8 cm long :

flowers orange red, 4 to 6 cm in diameter; calyx companulate with 5 to 7 lobes; petals; 5 to 7, wrinkled; stamens numerous ; ovary inferior, 3 to 7 celled ; fruit a modified berry, known as balausta, leathery skinned, spherical, with persistent berry, known as balausta, leathery skinned, spherical, 5 to 12 cm in diameter, brownish yellow to red, surmounted with persistent calyx; seeds numerous surmounted by pink juicy pulp. This acid pulp known to be 'aril' is edible portion. Double flower type is also known but they have only ornamental beauty.

Composition and Uses

Pomegranate is largely used as a dessert. The fruits are good source of sugars (14-16 per cent), minerals (0.7-1.0 per cent) and fair source of iron (0.30-0.7 mg/100 g) but poor in calcium. The quality of juice is predominantly judged by its sugar and content. The reducing sugars generally vary from 708 to 13.7 g/100 ml and the acidity from 0.45 to 3.47 g/100. Deciduous varieties are generally more acid than evergreen.

Pomegranate is used as a salad or table fruit and in beverages. The fresh fruit is of exquisite quality while its processed products like bottled juice, syrups and jelly are highly appreciated. The juice is considered useful for patients suffering from leprosy. The roots, rind seeds are important medicinally especially against diarrhoea. The juice is useful for leprosy patients. Dried seeds with pulp are available as 'Anardana'.

Climate and Soil

It thrives best under hot dry summer and cold winter provided irrigation facilities are available. Humidity lowers the quality and proliferates diseases. Under subtropics it is deciduous in nature while in topics it is evergreen. When dormant it can tolerate frost to considerable extent (-11.7°C) but new growth is damaged by late frost.

It grows under wide variety of soil and can tolerate even alkalinity and salinity to certain extent. However, best results are obtained in deep, heavy loan and well drained soils. It is sensitive to soil moisture fluctuation causing fruit cracking which is a serious problem of this crop.

Important Varieties

Until recently most plantation of pomegranate in India has been raised through seedling, therefore, well established varieties are rare. However in recent years this crop has received considerable importance and some recognized and descript varieties is coming up.

- **Ganesh:** This is a selection from 'Alandi' developed by Dr. Cheema, at Pine, which has revolutionized cultivation of pomegranate in Maharashtra state. This has soft seeds with pinkish flesh with juice of agreeable taste and bears heavily.

☆ **Alandi or Vadki-** It is commonly grown in Maharashtra (Pune) and in Gujarat. Fruits are of medium size. Blood red or deep pink flesh with sweet slightly acidic juice. The seeds are very hard.

☆ **Dholk:** Popular variety of Gujarat with fruits of large size. rind greenish yellow, fleshy aril pinkish white; seeds soft; juice acidic. It is a medium cropper.

☆ **Kandhari-** It produces large fruit. The rind is deep red and the flesh is a dark red or deep pink. The juice is slightly acidic. The seeds are hard. The variety is successfully grown in Himachal Pradesh.

☆ **Muscat-** It produces small to medium size fruit with thick rind. The juice is sweet, and the seeds are not very hard. It is commercially grown in the Kolhar, Rahuri region of Ahmednagar district of Maharashtra. High yielding and better quality strains have been selected from cultivar Mascat *viz.*, P-13, P-16, P-23, SK-1.

☆ **Paper Shell-** It is grown in South India. The fruits are medium to large size. The flesh is pinkish and the seeds are soft. The aril has good flavour. Fruit bearing capacity is high.

☆ **Kabul-** It also produces large fruit. The rind is thick deep red mixed with pale yellow. The flesh is dark red and the juice is slightly bitter.

☆ **Nabha-** Tree is medium in growth, fruit skin yellowish and smooth, cracking medium. Juice is 65 per cent. The acils are light pink in colour; seeds medium soft; taste is good.

☆ **Jodhpur Local:** Fruit medium to large in size; hard rind; fleshy aril light pink, sweet, juicy; seed moderately hard.

☆ **Seedless (Bedana):** Fruit medium to large in size; rind brownish green, fleshy aril light pink or whitish, very sweet, very juicy, soft seeded.

'Bassein seedless and 'KVK-1' are popular variety in Karnataka. Paper Shell','Kandhari' and 'Jalore Selection' are popular names in North India. In South India 'Paper Shell', 'Spanish Ruby', 'Muscat Red' and 'Velladu' have shown promise. Thus the efforts are on.

Hybrid Variety

☆ **Amlidana-** It is a cross between Ganesh x Nanha cultivar. It is dwarf and suitable for high density planting (HDP). Released from IIHR, Bangalore.

Propagation

As already stated, many pomegranate plantations in India have been raised from seedlings but it brings variability and therefore, vegetative propagation is a must for uniformity. Propagation by cutting is done commonly and there seems to be no difficulty in this method. Certain variety rooted when treated with 2000 ppm IBA for 30 seconds and when hard wood cutting were placed under mist. If the cuttings are treated with 200 ppm IAA, not only a majority of cuttings rooted but they also formed profuse root system.

The pomegranate is also propagated by air layering (goottee). Treatment with IBA 10,000 ppm in lanolin was better. Cuttings should be taken from suckers which spring from the base of the main stem and should be mature about 20 to 30 cm long band 6 to 12 mm thick. Rainy season is the best period to achieve maximum success.

Planting

The land selected should be thoroughly and deeply ploughed and harrowed. Planting place should be marked and pits of 60 x 60 x 60 cm are made well in advance and filled back with a mixture of FYM and soil. In subtropical climate, where plants remain dormant in winter, planting should be done in the beginning of spring season (February/March). However, in places where the plant behaves evergreen the proper time of planting is beginning of monsoons, when plants are lifted from nursery for transplanting in field, top should be pruned lightly to strike a balance with root system.

Planting distance would depend on soil type and climate; old orchards are planted at 3 to 6 m space which is smaller. The appropriate distance would be 6x6 m for permitting sufficient space for canopy and aeration; this is also the practice in rich alluvial soils of Gujarat and North India. However, in a spacing trial at MPKV, Rahuri, with 5x2, 5x3, 5x4 and 5x5 m spacing, it was observed that 5x2 m spacing gave maximum production. It was therefore, recommended that initially pomegranate should be planted at closer distance to obtain higher yields and if need be they can be thinned out. At 5x2 m spacing a crop load of 50 fruits per tree was recommended.

Training and Pruning

Pomegranate plant can be trained on a single stem of in multistep system. In kitchen gardens it is kept as a bush but for commercial orcharding it is uneconomical. In our country multistep training is more prevalent and useful also because due to some reason even if the plant loses one stem it continues to survive. At Rahuri varying of stem number from 1 to 5 did not affect the yield in initial years. However, a well-trained plant is an asset and it requires to be standardized.

Pomegranate fruits are borne terminally on short spurs attiring from mature shoots. These bear fruits for 3 to 4 years. Only a limited pruning of trees is required. The gradual growth of new shoots should be encouraged by restricted cutting back of bearing shoots. Suckers are removed regularly.

Intercropping

Since pomegranate plants take 4 to 5 years to come into good bearing, intercropping can be beneficial for which low growing vegetables, pulses green manure crops may be grown depending on the resources and climate. Under arid, they are recommended only during rainy season.

There are many other fungal spp. which are associated with fruit rot in storage like *Glomerella cingulata*, Botrytis *cinerea*, *Aspergillus castaneces*, *Phomopsis spp.*, *A. niger* and *A. foetidus*. All these can be kept under control by spraying Diathane M-45 or Captan as stated above.

Physiological Disorder

Fruit cracking: Fruit cracking is a most serious physiological disorder in pomegranate which limits its cultivation. In young fruits, it could be due to boron deficiency but fully grown fruits crack due to moisture imbalances as this fruit is very sensitive to variation in soil moisture and day and night atmospheric moisture deficit. Prolonged drought causes hardening of peel and if this is followed by heavy irrigation or down pour then the pulp grows and the peel cracks. There are some varieties which are tolerant to this disorder like 'Bedana Bosec' and 'Khog'. This problem can be managed through following means:

a. Maintaining soil moisture and not allowing wide variation in soil moisture depletion.

b. Cultivation of tolerant types.

c. Early harvesting, not allowing fruits to crack.

d. Spray of calcium hydroxide on leaves and fruits initiating after fruit set

Harvest and Yield

The fruits are ready for harvest 5 to 6 months after anthesis depending on the season of the crop. They should be harvested mature, which can be judged by change in skin colour, which turns slightly yellow and fruit looks fully developed and gives metallic sound when tapped. Plucking is done with hand. A grown up well kept tree should give about 100 to 150 fruits per annum and such tree continues to give economic crop up to about 25 to 30 years. A high density system with 'Ganesh' 1000 plants/ha (5×2 m) has been found very profitable giving a crop load of 50 fruits/tree.

Chapter 27

Pilu

Botanical name:	*Salvadora oleoides*
Family:	Salvadoraceae
Chromosome no.:	2n= 24
Origin:	India

Introduction

A very common tree found growing luxuriously in Indian deserts especially in Rajasthan and Gujarat. It is generally found growing in highly saline region of desert and makes and excellent food for camel during summer. Fruits is eaten by natives and seeds are rich in oil (40-50 per cent) which is non edible but have lot of industrial value and therefore attracting the industrial attention. It is an excellent plant for shelter belt and wind break in desert as it make almost impenetrable growth through root sucker around the stem and seedling. Which spring up under the shade of canopy.

Taxonomy and Morphology

Pilu belong to family salvadoraceae and genus salvadora of which two species *Salvadora oleoides* Decene and *Salvadora persica* L. are commonly found in India. Plant is an evergreen shrubs or small tree with short twisted trunk and druping branches leaves opposite, liniar, lensiolet, entire rather thick. Flower sessile greenish white in peniculate spike, terminal or auxiliary. Fruits aromatic drupe, globose, usually yellow when ripe.

Composition and Uses

Fruits are sweets and usually enjoy by native. But the leaves make excellent food for animal. Seeds are 44-46 per cent of total fruit which contain 2.8 per cent moisture, 45-48 per cent fat, 18.94 per cent albuminoids, 23.48 per cent carbohydrate, 5.8 per cent fiber and 3.5 ash. Fruit pulp is rich in glucose, fructose and sucrose. Purified oil can be use for soap making and can substitute coconut oil can be use for candle, rheumatic pain and as a base for ointments and oil cake can be used as animal food. Fruit is useful in a large spleen, rheumatism and low fever.

Climate and Soil

It is most suitable for arid climates of tropics and subtropics. It is sensitive to frost.

It requires sandy to sandy loam soil. It can tolerate salinity to a great extent.

Important Varieties

There is no descript varieties are available however it has great natural variation is available. The plant with fruit with higher sweetness higher oil content which lower level of compound like diebenzile thio urea, dibenzile urea and thio-glucocide, gluco troplolin which make its oil non edible should be selected.

Propagation

The plant occupies fairly well and regenerates through root sucker and to some extent through natural layer. Seeds can also be use for regeneration.

Planting

One year old plant can be used for planting. In the beginning of monsoon so that plant get sufficient time to establish. Once established they would not need any special care. Looking to the size that plant attains naturally and amenability to lopping, planting could be at (6×6 m) for which pith of (60×60×60 m) in summer refiled with mixture of compost and soil (1:1).

Nutritional Management

A fully grown, bearing plant should be top dressed with fertilizers @ 100 g N, 160 g P_2O_5 and 260 g K_2O. These may be applied in two splits, when there is sufficient moisture in the soil.

Irrigation

Young plants need to be cared for watering time to time in the first year of planting and when plants establish well, usually watering is not done. However, sufficient moisture helps in good yields for which runoff collection should be practiced in rainy season by developing micro catchments around the trees.

Flowering and Fruiting

A plant shall take about 5 year to begin bearing but it will be determinded by the method of propagation and management practices. Flowering continue for a long period from January to April and the fruits are thus available for a protracted period of 3 month from April to June. The production is about 5-10 kg per tree the ripe fruit turn yellow when they are picked by hand.

Insect Pest and Disease

There is no insect pest and disease which are damaged economically. During the initial stage of establishment, termites attack the plants. The control termite drenching with chloropyrophos at 2 ml/lt water or aldrin dust 500 g/pit should be applied. Both seedling and mature plants are affected by phllody or witches broom disease caused by phytoplasm viral and crown gall diseases. It can be control by spray of Bavistin at 2 per cent.

Harvesting and Yield

Fruits are harvested under protracted period of 3 month from April to June. The production is about 5-10 kg per tree the ripe fruit turn yellow when they are picked by hand.

Chapter 30

Starfruit (Carambola)

Botanical name:	*Averrhoa carambola*
Family:	Oxalidaceae
Chromosome no.:	2n = 24
Origin:	Malaysia or Indonesia

Introduction

It is an important minor fruit in tropical regions of the world. The tree said to be native of Malaysia, but now it is cultivated throughout India for edible fruits which are mostly used as substitute of tamarind while cooking. It is one of the important and popular minor fruit in China, Taiwan, Malaysia, Thailand, Pakistan, and Indonesia. Carambola is a small tree growing to bushy stature and sometimes pruned to in penetrable hedges. In Indian languages it is known as Karmal, Kamrakh and Kamranga. In India carambola grow primarily in the southern states and along the west coast, extending from Kerala up to West Bengal.

Taxonomy and Morphology

Carambola belong to the genus of *Averrhoa* along with its two other sister species of *bilimbi* and *acida* in the family of Oxalidaceae. The height of plants goes up-to 10 meters with criss-cross branches spreading closely to give the plants a bushy shape. It is an evergreen plant producing hermaphrodite white flowers marked pink and shining oval leaves produced in pair of 5-7 on the twig. The members of genus *Averrhoa bilimbi* known as cucumber tree and *A. acida* produces small and more acidic fruits in cluster.

Composition and Uses

The fruits are acutely angled but fleshy and succulent. Unripe fruit are acidic due to presence of oxalic acid (0.96 per cent) and malic acid (0.06 per cent). On ripening, the acidity reduces gradually and these come down to only 0.22 per cent. Fruits are also good source of vitamin A, B and C, minerals specially Fe, K, Mg and P. It is used as a dessert fruit or for preparing juice, jam, jelly and pickles. It has medicinal value also. Ground leaves and shoots for chicken pox, root extract as antidote for poisoning and pulp of immature fruits for cleaning brassware. It is used as an alternative to tamarind in cooking.

Climate and Soil

Carambola grows well in both tropical and subtropical climate free from frost. But it prefers warm and moist conditions optimum temperature range is 21-32 °C. In addition light to heavy rainfall throughout the year is preferred. In South India It is seen to grown on an altitude of 1200 meters.

It can be grown in any type of soil. But well drained, deep clay soils are preferable optimum p^H is 5.2 to 6.2. Calcarious soils are not suitable. It cannot with stand water logging.

Important Varieties

No specific cultivars but 2 types namely sour and sweet are grown in India. Sour type (1 per cent acid) and Sweet type (0.4 per cent acid).

- ✰ **Florida-** Golden Star, Fwang Tung, Dah Pon, Tean Ma, Arkin, Maha, Star King. Thai Knight, Wheeler, New Comb.
- ✰ **Singapore-** Leng Bak, Jurong.
- ✰ **Taiwan-** Lu Tho, Cheng.
- ✰ **Malaysia-** B_1, B_2, B_4, B_6, B_8, B_{10}, B_{11}.
- ✰ **Australia-** Giant Siam, Jungle Gold, BCP-1, Hosic and Chujuba.
- ✰ **Thailand**- Fwang Tung, Thai Knagut.

Propagation

Plants are mostly propagated by seeds. Asexual method of propagation by grafting, budding and layering is also recommended. Approach grafting is, however, very successful. M-18960 as stock is best for calcareous soils. Seeds should be sown immediately as they loose viability at a faster rate. Micro propagation is also being attempted.

Nutritional Management

Carambola is a grass feeder and fruit size is increased with good supply of nutrition and watering. The rate of mixture 10 kg N, 2 kg P_2O_5 and 17 kg K_2O per tree are recommended. Addition of 'N' fertilizer improves the yield. Normally organic manures are only applied.

Irrigation

Irrigations are given during fruit development only if the weather is dry. Trees may be provided 200 litre/tree/weak in dry periods. Mulching will be very helpful in reducing weed growth and irrigation needs.

Planting

The planting distance varies from 7-9m x 5-7m or high density planting with 7x3m or 4.2x3.6m is also practiced in some countries depending upon the cultivar. One year old healthy layering or grafted plants are planted in prepared pits of 60x60x60 cm. The pits are filled with organic matter and left for 2 months before planting. Planting can be done throughout the year but monsoon season is best for good establishment.

Training and Pruning

Plant should be trained to have four to six main scaffold limbs from suitable height on the trunk. Pruning back old branches will improve quantity and quality of the fruit.

Flowering and Fruiting

The flowers are borne laterally on the mature shoots or trunks of the tree in cluster in small. Flowering in carambola is found in three flushes-spring flush (April-May), rainy flush (July-August) and autumn flush (September-October) like other tropical fruit plants. Fruits from the above flushes could be harvested in July-September, November-December or February-March respectively.

The rate of fruit growth was slightly rapid in first 3 weeks after fertilization when about 25-27 per cent growth was complete. The total period required from flowering to ripening of fruit from 100 to 110 days in the cultivars under sour and sweet.

Insect Pest and Diseases

Fruit fly, borers, fruit piercing growth, aphids, mealy bugs, scales. Leaf spot and anthracnose- any fungicide can be used.

Harvesting and Yield

Carambola produces fruit on both young and old branches and even on trunks in clusters. Grafts start bearing within a year but good yields can be had after 3 years. Seedlings take 4-5 years to come to bearing. Carambola being non-climacteric should be harvested at proper stage of maturity only, *i.e.* when 50 per cent of fruit becomes yellow. On ripening fruit develops orange tinge. Carambola produces fruit throughout the year but peak seasons of harvest are January-February and September-October. Yield varies from 20-50kg/tree in the early stages and a grown up tree may give 400-500 kg/tree.

Chapter 31

Tamarind

Botanical name:	*Tamarindus indica* L.
Family:	Leguminocae
Chromosome no.:	2n = 24
Origin:	Tropical Africa

Introduction

It is one of the most popular tree found cultivated throughout India as stray plantation or avenue. Regular orcharding is wanting but common in waste land. Therefore correct estimate of total acreage under this species is difficult. The wood is used in card wheels, rice pounders, oil mills *etc.* The bark is used in tanning, the tender leaves and flowers are eaten as a vegetable. The decoction of bark is useful in diarrhea. The pulp of fruit is used for culinary purpose. The seeds are rich in pectin.

Taxonomy and Morphology

Genus *Tamarindus* has only one species *indica*, belongs to the family Leguminocae. Tree is large, 40 to 60 feet high; dark brownish or dark grey; leaves equally pinnate, 20 to 40 opposite leaflets, close, oblong, blunt and smooth. Flowers yellow stripped red, only the upper 3 in racemes are developed; racemes are a few flowered. Fruit is a pod 3 to 6 inches long and 1 inch wide, 3 to 10 seeded, shaped. Pulp around the seed is eaten.

Composition and Uses

Fruit pulp is contain of 55.0 per cent of pod, contains about 20.6 per cent water, 3.1 per cent protein, 0.4 per cent fat, 70.8 per cent carbohydrates, 3.0 fibers, 2.1

ash and 10 per cent acidity due to tartaric acid. Seeds have 63 per cent starch, 16 per cent protein and 5.5 per cent semi-drying oil.

This tree is generally grown for its fruits which are used either mature or. Immature fruits are used for chutney and ripe fruit pulp is an important ingredient in South Indian vegetable dishes to induce sourness. These fruits also make excellent chutney. It is believed that continuous use of tamarind in daily food reduces the chances of stone formation in urinary system of human beings. It is an important condiment. Pulp is mixed with sugar and made into tamarind balls (sweetmeat). It is used for seasoning other food and in curries, preserves, chutneys and sauces. A refreshing acid drink and syrup are also made. Seeds are eaten roasted or boiled. Used as source of carbohydrate for sizing cloth, paper and jute products.

Climate and Soil

Semi-arid tropical and sub-tropical region but can be grown in monsoon region with good drainage. Fruits so not ripe in cold and is sensitive to frost. It prefers warm climate.

It is sown in forest and waste lands in arid region it can be grow in poor soils. A compost of fibrous loam and sand suits very much.

Important Varieties

Since cultivation has been through seeds no standard cultivars are available. However, some selections are reported from Rahuri (Maharashtra) and regular plantation is being initiated in Maharashtra and Tamil Nadu. Great variation in available from sweet to sour type and they require exploitation after evaluation.

Propagation

Seed propagation is common. Budding is also successful. Cuttings also root easily during rainy season or in glass house. The most appropriate technique for arid region is in situ soft wood grafting on one year old seedling when new growth takes place for which activated scion should be employed. Summer is the appropriate season when new growth takes place.

Planting

Tamarind tree attains a huge size; therefore it should be planted at 12×12 m in a field or 10m apart as avenue or farm boundary. For this pits of 1x1x1 m size are prepared in summer, refilled with soil + 50 kg compost and planting should be done in the beginning of monsoon. It would be desirable to develop micro-catchment around trees to collect runoff water from monsoon rains.

Training and Pruning

Permit clear high head and uniform branching in all direction for the natural growth of the tree. No pruning is required.

Intercropping

This is a slow growing tree and would take about 15 year to occupy allotted space at 12×12m. Under these conditions during rainy season green manure crops or appropriate vegetables can be taken to economise early years of orcharding. Choice of the crop will be determined by the climate of the region.

Orchard Soil Management

Basins should be kept clean to permit accumulation of water in rainy season and interspaces could be covered with appropriate vegetation to avoid soil erosion. In summer, mulching would help to avoid water loss through evaporation. However, these standards will have to be investigated. Presently plants receive no attention.

Nutritional Management

Being a leguminous tree, may not require nitrogenous fertilizer but permission of rooting of leaves which shed regularly in the drip shall be beneficial in enriching the soil and improving its physical state. Phosphorus application at pit filling will be advantageous.

Irrigation

Once the plant has established its hardly needs any irrigation However, water harvesting in rhizosphere during rainy season would be advantageous. In heavy rainfall areas proper drainage would be essential.

Flowering and Fruiting

In tamarind foliage changes in March to April, flowering takes place in May to June and the fruits are available in February to March. A seedling tree will take 13 to 14 years for first flowering but a vegetatively propagated plant in precious and comes into bearing in 7 to 10 years. As tree grows in size and age, productivity increases and continues to be productive for more than 60 years.

Insect Pest and Diseases

Several insects attack the fruits and seeds of tamarind both on the tree and when in storage. These include: the beetles and Larvae of *Lasioderma serricorne, Pachymerus gonagra*, and *Calandra linearis. Aspidiotus spp.* feeds on fruit sap and damage.

The insect can be control by systemic insecticide like rogar or Monocrotophas or Diemecron or Dimethoate 0.05 per cent.

The tree is affected by a number of rots, such as saprot (*Xylaria euglossa*), brownish saprot (*Polyporus calcuttensis*) and white rot (*Trametes floccose*).

The disease can be control by fungicide like Dithane Z-78 or Bavistin 0.2 per cent solution.

Harvesting and Yield

Fruits generally mature in February to March in which exocarp gets hardened and separated from the pulp. Fruits are plucked individually or shaken off and collected. When harvested the fruits are almost Semi-dried and do not spoil due to high acidity. A fully developed tree can give production of 200 to 250 kg per annum.

Chapter 32

Wood Apple

Botanical name: *Feronia limonia*

Family: Rutaceae

Chromosome no.: 2n = 18

Origin: India and Ceylon

Introduction

Wood-apple, it may be called elephant apple, monkey fruit, curd fruit, *kaith bel* and other dialectal names in India. The slow-growing tree is erect, with a few upward-reaching branches bending outward near the summit where they are subdivided into slender branchlets drooping at the tips. The bark is rigid, fissured and scaly and there are sharp spines: 3/4 to 2 in (2-5 cm) long on some of the zigzag twigs. It is one of the very hardy tree grown in semi arid and arid region of the country. It is popular tree of indoor yard garden and can be exploited for regular cultivation in dry areas.

Taxonomy and Morphology

It is a member of Rutaceae family. It is a spiny deciduous tree with odd pinnate leaves 3-7 leaflets. The surface of fruits and branches are very rough and covered with white bloom, the flowers have dull red color and appear in terminal or auxiliary panicles. Fruits is hard shelled many seeded berry and contain, at maturity, a single cell filled with pinkish edible pulp in which seeds are imbedded.

Composition and Uses

The rind must be cracked with a hammer. The ripe fruits contain sour sweet aromatic pulp, which is about 70 per cent of total weight, in which seeds are imbedded. It contains about 70 per cent moisture 7.3 per cent protein, 0.6 per cent fat and 1.9 mineral matter. Fruits are rich source of (riboflavin 170 mg per 100 gm), acidity is about 2.3 per cent and sugar 7.2 per cent the pulp is used for preparing chutney and due to rich in pectin it makes good jelly but astringent and therefore can be mixed with guava jelly. The pulp represents 36 per cent of the whole fruit. The pulp can be freeze-dried for future use but it has not been satisfactorily dried by other methods.

Climate and Soil

Wood apple is mostly grown in dry region of sub-tropical and tropical zone of the country. It grows well above 1500 m see level. It is being a hardy plant can sustain in low rainfall area covering semi-arid or arid climatic condition. It is seen to grow appreciably within a temperature range of 0°C to 48°C but optimal temperature for adequate growth and productivity ranges between 25°C- 30°C.

It can be grown in wide variety of soil including degraded soils of arid region and can tolerate salinity to certain extent. It can also be grown in waste land.

Important Varieties

There is no descript variety are available however great variation exist which could be surveyed and exploited. Some sour and sweet type has been reported by high productivity and big fruited.

Propagation

Generally wood apple plant is propagated by seed however budding is successful and can be done in late summer or early monsoon. The budded plants are dwarf precious and prolific in flowering and fruiting. It would be advantageous to go for in situ budding under arid condition.

Planting

Organized cultivation in the form of wood-apple orchard is not seen. However, to establish a wood apple orchard or to make a solo planting, soil should be worked and pits measuring 1x1x1 m should be dug during summer. In arid condition planting should be done in the begging of monsoon during the month of July to September the planting distance should be kept at (8×8 m) at the time of planting basin should be prepare for collection of runoff water around the tree to survive the period subsequent to rainy season.

Training and Pruning

Trees are allowed to grow freely on central leader with well spaced branches in all direction. The tree is deciduous in nature but no pruning is required.

Nutritional Management

It is very hardy crop there is no need of manuring but it will be benefited if manured at the rate of about 25 kg per tree in the beginning of monsoon. It will help in increasing the fruit size and quality.

Irrigation

Wood-apple is normally a rainfed crop as such no irrigation schedule is followed to cultivate the fruit tree. It is a crop of dry region and once the plants had established they hardly need any irrigation. Subsequently, one or two irrigation during drier months between Nov.-March will facilitate better canopy development in pre-bearing stage. Once the tree start flowering, one irrigation after fruit set and a second irrigation during fruit development, particularly second phase of development, will help better size and more retentions of fruits. Nevertheless conservation of runoff rain water in rhizosphere will enhance the productivity of this crop.

Flowering and Fruiting

The period of flowering and fruiting is governed by climate and the period of moisture availability, it flowers from February to May and the fruits are available in winter. A well grown tree bears 200-250 fruit per annum.

Insect Pest and Diseases

Since this plant is not under regular orcharding therefore there is no record of any serious disease and pest. Insect and disease which affect the citrus spp. may cause damage to this crop also.

Harvesting and Yield

A vegetatively propagated tree start bearing at three (3) year after transplanting however it take about 10 year for a tree to provide optimum production. Generally fruits are ready for harvest in November onwards during winter and the stage of harvest is determined by the use of fruits. For chutney immature but fully develop fruits are preferred. However for squash fully mature fruits are desirable. A well grown tree yielded 200-250 fruit in a year.

Index

T

V

W

X

Z

9 789359 191157

Printed at Repro India Ltd.